A Rolling Die Gathers No Data

Lab Activities for Statistics

Eric Kamischke
Ellen Kamischke

Natural Log Enterprises

Cover design by Rachel Kamischke

The Authors

Eric and Ellen Kamischke have been teaching mathematics for a combined total of 70 years. They have taught both high school and college level students in statistics and most of the courses that precede and follow. They are co-authors of both Discovering Algebra and Discovering Advanced Algebra published now by Kendall Hunt. Eric has been a reader for AP Statistics, and Ellen a reader for AP Calculus for over ten years.

Acknowledgements

Our thanks first go to the many years of students who have done these labs, written their reports, and provided us with feedback about what they learned (and didn't learn). Secondly we thank Key Curriculum Press who willingly published our first book of activities when we were much younger.

Limited Reproduction Permission

Your Help

Eric and Ellen are responsible for the editing, the mathematics, the layout, the drawings, and tables. We would appreciate any help from you, our readers, or your students if you should find any errors that we may have missed. Please contact us at the website below with any feedback that might improve this book.

Proceeds from the sale of this book support the New Life Primary School in Mizak, Haiti.

Natural Log Enterprises
1220 Reads Run
Traverse City, Michigan 49685
www.naturalloge.com

ISBN 978-0-9962372-1-5

Contents

Note to Teachers

The lab activities in this book were developed over a number of years while teaching AP statistics and a first course in statistics in college. Just which ones are used and how they are used will depend upon the course and your students' ability levels. A topic chart is included to help identify where each lab activity fits in the curriculum. In cases where more than one topic is noted, the investigation seeks to tie concepts together and enable students to see the connections between various ideas.

Some of these investigations are designed to enable students to discover a new method or concept. Others are designed to clarify some confusing point for students. Most of the labs are designed to be done in small groups, while others work best for students working in pairs or as projects for individuals. These distinctions are noted in the grouping chart following these remarks. We usually allow students one to two weeks to complete an activity working outside of class. The activities take considerably less time if students work on them in class. We have generally required three or four investigations per semester for each group of students. In addition, some of the shorter investigations have been used as extra credit options, class investigations, or even in-class assessments. Lab 21: The Project has been used as a full length activity starting the first week of class and ending the last week of class.

Data Collection

Many of the labs begin with data collection. You can shorten these activities by providing your students with sample data. While this may be necessary at times to fit things in, we have found that students working with data they have collected, enhances learning and understanding. Given a canned data set students always suspect that this data was carefully groomed to make it work. We can assure you that all the sample data sets given in this book are not groomed, but actually collected and in raw form. But if you want your students to see that it is not the data that makes it work, you need to let them collect their own. A brief equipment list is included. In most cases there are alternatives when equipment is needed, so a lack of equipment should not preclude a class from doing the investigation. Check the teacher notes for each lab for potential alternative equipment options.

Simulations

There are many good simulation activities related to probability, sampling distributions, confidence interval, and testing. We have resisted putting any computer or calculator simulation activities in this book. This is because (1) there are already good sources for these available and (2) we have found that some students, no matter how much explanation is provided, just do not understand what is being simulated and what they can learn from it. The simulations you will find involve as much human intervention as we could include to help the students know where things are coming from.

Student Lab Reports

You may want to copy the page entitled Student Instructions for Writing a Standard Lab Report, although you may want to modify these instructions to suit your own particular specifications. Reports for various activities may differ. Activities that are very structured should take the form of summarizing what is done and learned in each step. Expect students to write the answers in complete sentences. Encourage students to explain what was learned in the step and not just give the answers to the questions. In more open ended activities, students should include detailed accounts of their procedures, even for those attempts that were not successful. By describing and analyzing what went wrong, students gain additional insight into problem-solving. Students should annotate all calculations and equation work. They are learning to write for an educated audience. The need not explain every detail, but they should use proper vocabulary and express their ideas clearly. You should be able to follow their thinking and calculations without having to fill in too many gaps.

Group Evaluation Form

The group evaluation form provides an excellent opportunity for you to get some feedback on how the groups are working. This is a confidential report from the students to the teacher. Be sure students understand that as members of a group they are expected to show an effort in working on the lab, participate with their group as much as they can, be cooperative and helpful to the others, be accessible to the rest of the group, and communicate their findings effectively. Students should feel free to vent frustrations with other group members or with the activity itself in this document. They should be encouraged to complete the evaluation thoughtfully. To provide this encouragement, you can award points for turning in a thoughtful evaluation for each lab. Reading the evaluations from various group members gives a picture of just went on as they pursued the activity. Sometimes students within the same group will have very different assessments of just how much effort each person put in. If there is a consensus that some individual is not pulling her or his weight, it is a good idea to speak with each member individually to get a clearer picture of exactly what the problems are. Then it is time to have a chat with the whole group. Without pointing fingers at the individual, you can discuss the responsibilities of each member of the group and how to make a successful effort on the next project. You may also want to have that group check in more frequently and report on what each individual is doing. It is also entirely appropriate to adjust individual scores on a project based on input from the evaluations.

Summary

Incorporating lab activities into a course takes time. However, the gains in understanding and appreciation of the power and usefulness of statistics are well worth the effort and time spent. We hope that you and your students will find these activities challenging and rewarding.

Eric and Ellen Kamischke

Topic Chart	Graphical Displays	Statistic formula	Probability	Distributions	Data Collection	Sampling Distribution	Linear modeling	Confidence Intervals	Hypothesis Testing
Puzzling Pies	✓								
Bin There, Done That	✓								
Finding Your Center		✓							
Spread it On		✓							
Thinking Outside the Boxplot	✓	✓							
Getting the Drop on Association	✓	✓					✓		
Throw Me a Curve, Bell			✓	✓					
The Apple and the Tree	✓			✓	✓				
Is This Normal or Not?	✓			✓					
By Chance or By Choice	✓				✓				
Census Sensibility					✓				
The Inch Pinch		✓			✓				
A Rolling Die Gathers No Data			✓	✓		✓			
It's in the Cards						✓			
The Confidence Game								✓	
p is for Pepper				✓					1T
Insignificantly Significant									1T, 1P
A Safe Prediction							✓	✓	
Is That Die Still Rolling?									1P, GOF
Jumping *F*rogs and *T* for Two									2T, ANOVA
The Project	✓	✓			✓		✓		2T, Slope

New Life School

In the small rural community of Mizak, Haiti, a group of young artists created Living Media International, a non-profit organization seeking to bring beauty and joy to their region. One thrust of this was to build and run a school that accepts all students and provides a unique quality education to those children. Ellen and Eric have supported this effort since its beginning and gladly give the proceeds of this book to support the teachers and administrator of this school. For more information on the New Life Primary School visit http://livingmedia.org/programs/new-life-primary/

Grouping Chart	Individual	Groups of 2 or 3	Groups of 3 or 4	Groups of 4 to 6	One hour or less	Up to one week	More than one week	Full Semester	Includes Data Collection
Puzzling Pies				✓			✓		✓
Bin There, Done That			✓			✓			
Finding Your Center	✓	✓			✓	✓			
Spread it On		✓	✓			✓			
Thinking Outside the Boxplot			✓			✓			
Getting the Drop on Association			✓			✓			✓
Throw Me a Curve, Bell		✓	✓			✓			✓
The Apple and the Tree			✓			✓			✓
Is This Normal or Not?			✓			✓			✓
By Chance or By Choice		✓	✓			✓			
Census Sensibility		✓	✓		✓	✓			
The Inch Pinch			✓				✓		✓
A Rolling Die Gathers No Data			✓			✓			✓
It's in the Cards				✓		✓			✓
The Confidence Game				✓	✓	✓			✓
p is for Pepper	✓	✓			✓				
Insignificantly Significant		✓	✓			✓			
A Safe Prediction			✓			✓	✓		✓
Is That Die Still Rolling?			✓			✓			✓
Jumping Frogs and T for Two			✓				✓		✓
The Project			✓					✓	✓

New Life School

In the small rural community of Mizak, Haiti, a group of young artists created Living Media International, a non-profit organization seeking to bring beauty and joy to their region. One thrust of this was to build and run a school that accepts all students and provides a unique quality education to those children. Ellen and Eric have supported this effort since its beginning and gladly give the proceeds of this book to support the teachers and administrator of this school. For more information on the New Life Primary School visit http://livingmedia.org/programs/new-life-primary/

Lab Equipment Needed

The labs here are written as done in our classes. It is assumed that students will have access to computational and graphing technology for every lab. There are many substitutions to the equipment listed here. Some of these are discussed in the teacher notes as well as other optional equipment to help labs go smoothly. In most cases sample data sets have been provided for when time is short. Using these will negate most equipment needs. Please do not reject a lab based on this table until you have read the teacher notes.

Lab	Equipment
Puzzling Pies	Protractor
Bin There, Done That	
Finding Your Center	
Spread it On	
Thinking Outside the Boxplot	
Getting the Drop on Association	Rubber band, water bottle, meter stick
Throw Me a Curve, Bell	Wire, ruler
The Apple and the Tree	Pennies, meter stick
Is This Normal or Not?	Stop watch, Video clip
By Chance or By Choice	
Census Sensibility	
The Inch Pinch	Ruler, tape
A Rolling Die Gathers No Data	Foam die, knife
It's in the Cards	Deck of cards
The Confidence Game	Deck of cards
p is for Pepper	
Insignificantly Significant	
A Safe Prediction	Rubber band, water bottle, meter stick
Is That Die Still Rolling?	Foam die, knife
Jumping Frogs and T for Two	Different paper weights, frog directions
The Project	As needed

Student Instructions for Writing a Standard Lab Report

The formats of the labs in this book vary greatly. Some of them are rather "cookbook" in structure. In some activities you're given a set of steps with questions to answer as you complete each step. Other activities are less structured, and you will have to define your own procedure and techniques. Some lab activities can be completed in an hour or two. Others will take more than a week to complete. However, your lab report for each activity should follow the guidelines given below.

- **Statement of Objectives:** For some projects you are given the objective. For others you have to determine why you are doing the lab and what the goals are. Make this section of the report concise and clear.
- **Data and Procedure:** The report should stand on its own. The instructor should not have to look at a copy of the project page in order to understand what is happening. For the less structured activities, include a complete explanation of what you did and the order you did things, including why you chose to do the lab in that way.
- **Calculations and Questions:** Be sure that calculations are clearly labeled and commented upon when they are not entirely self-explanatory. The questions asked in the lab are often the point where real learning can happen. Discuss each answer as a group. If the group agrees to disagree then make note in your report as to why you are not able to decide as a group.
- **Plots and Graphs:** Each graph should be titled and have the axes labeled with units indicated. Use appropriate scales. Graphs may be hand-drawn or printed from a computer or calculator. Use whatever technology presents the material most clearly. Most plots should be followed by a line or two of text indicating what important is shown in the plot.
- **Conclusions:** In this final section, summarize your results. What new discovery did you make? What relationship did you find? Where there any significant sources of error? If so, can you suggest improvements on the procedure to reduce these errors? If you are using actual measurements, be careful with units on your statistics.

Group Evaluation

Lab Project Title: ____________________

Group Members:

(In the box following each name, rate each member's overall performance in the group on a scale of 1 to 5. Consider their effort, participation, cooperation, accessibility, and communication skills.)
1 is a poor group member
2 is a weak group member
3 is an average group member
4 is a strong group member
5 is an exceptional group member

- Your Name ____________ ☐
- Other group members:
- ____________ ☐
- ____________ ☐
- ____________ ☐

Please respond to all of the following questions.

1. **Describe your contribution to the project.** Specifically, describe what you did to assist in the completion of the project questions.

2. Choose one statement that best characterizes your contribution to the group effort and explain on the back.
☐ I did more than my share. ☐ I did my share. ☐ I did not do my share.

3. **Choose one statement that best characterizes the contribution of each member to the group effort and explain.**
   - ☐ Each member of the group contributed equally to the completion of the project.
   - ☐ Other member(s) of my group contributed more than a fair share to the completion of the project. (Include name(s) of the group member(s) in question.)
   - ☐ Other member(s) of my group did not contribute a fair share to the completion of the project. (Include name(s) of the group member(s) in question.)

4. **Were there difficulties in scheduling times to meet?** Explain.

Lab 1
Puzzling Pies

Equipment needed

- Protractor

Part 1

Some students collected a set of data and created a nested pie graph. However, part of the graph was destroyed by an ink spill, and part of the data misplaced. The incomplete data and graph are shown below. This creates a puzzle to be solved. By studying and measuring the graph, recreate both the complete data table and the complete graph.
Variable 1: Lunch time. Students were classified as those who ate before 12 noon, and those who ate after 12 noon.
Variable 2: Food. Students were classified as those who purchased no food (brought their lunch from home), students that mixed some food from home with some purchased food, and students that purchased their entire meal.

Food purchase	Before noon	After Noon
None	15	24
Some	25	??
All	??	18

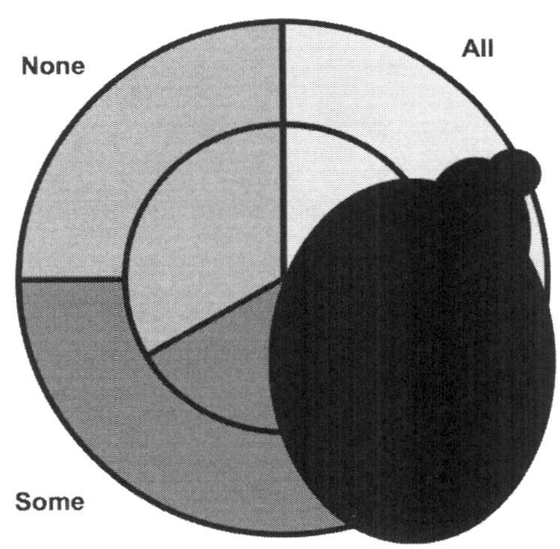

Outer: Before 12
Inner: After 12

Part 2

In Part 1 you solved a puzzle. In this part you must create your own similar puzzle for someone else to solve.

a. Begin by collecting categorical data on two variables as in the example. For your puzzle, one of your variables must have 3 values, and the other must have 4 values. (The variables in Part 1 had two and three values.) You may use human subjects as Part 1 did, or you may use other subjects such as books in the library, pencils and pens, or houses in a neighborhood. You will need between 100 and 200 subjects. Choose your two variables so that your values fit into 3 categories for one variable and 4 categories for the other variable. Collect the data.

b. Create a nested pie graph of your data set as in the example. Your graph will have 3 rings and four sections in each ring where the example had two rings and three sections in each ring.

c. Create your puzzle. Note that in Part 1 you could only complete the graph or the table by studying both of the sources. Your puzzle must require the same solution strategy. If the reader can solve the puzzle using only one of the two items, then it is not a good puzzle. Part 1 had two values missing from the table and two dividing lines missing from the graph. Your puzzle must have at least three values missing from the table and three lines missing from the graph.

d. Prepare one sheet of paper with your "incomplete" table and "incomplete graph" as the only things on that page.

Part 3

You report should include ...

a. Your solution to the puzzle in Part 1 (that is, the complete table and complete graph) and an explanation of how you solved it.

b. Your complete data table and complete pie graph for Part 2.

c. Your page with your puzzle created from the incomplete Part 2 data.

Teacher Notes for Puzzling Pies

Prerequisite Knowledge

- Reading a nested pie graph
- Constructing a pie graph
- Reading a two way table of data

Equipment needed

- Protractor

Objectives

The objective of this lab is to help students clearly understand the connections between data and a categorical graph of that data. Too often students use tools to create graphs without understanding how those graphs could be created by hand. Working through the puzzle helps students understand these connections.
This lab works best with groups of 4 to 6 students. Part 1 should take less than a half hour. The data collection in part 2 may take several hours, though if the collection duties are split among group members, each one should not have to spend more than a half hour collecting the data. Part 3 will take less than an hour. If the work is done in class, plan to spread the project over a week, allowing for two or three days between Part 1 and Part 3 with Part 2 done outside of class. Allowing 15 - 20 minutes in class for Part 1 and the same amount of time on 2 days for Part 3, should be sufficient.

Comments and Answers

Part 1

There are several approaches to this puzzle. If the group is large, then recommend they split in half and each work independently on the puzzle to compare later.

Food purchase	Before noon	After Noon
None	15	24
Some	25	**58**
All	**60**	18

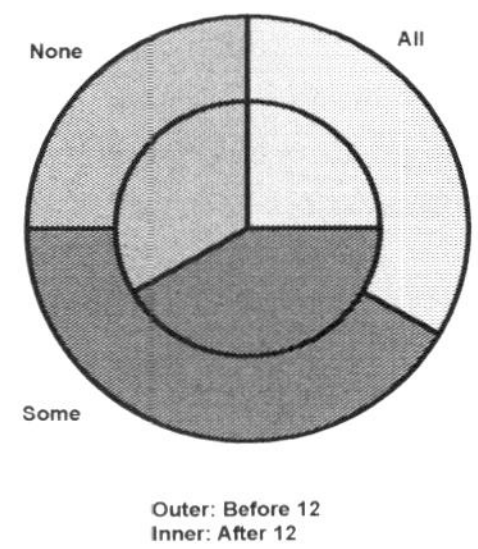

Part 2

This is not an activity on proper sampling. There is no need to include randomization or even care if they are sampling more than 10% of the population. However, it is still a good idea to have students verify their data collection plan with you before they actually collect the data. In this way you can be sure they have the correct number of values for each of the variables. Encourage the group to split the data collection duties among the group members.

Part 3

Encourage students to try to solve their own puzzles as they create them. Creating an accurate pie graph by hand may be challenging. Encourage students to explore using technology such as an Excel spreadsheet to assist in this process.

Part 4

As part of the assessment for this assignment, you might give photocopies of the puzzles to other groups to solve.

Lab 2
Bin There, Done That

For any data set there are an infinite number of histograms that can be drawn. The bin width (or number of bins) is the first decision to make, and the starting value for the first bin is the second decision.

Most electronic statistics tools will "auto fit" these two features for the user, but often technology does not choose well. With the "correct" number of bins, the reader can accurately estimate the mode, median, mean, and see the shape of the distribution (the whole population) at the same time as they see the values of the sample.

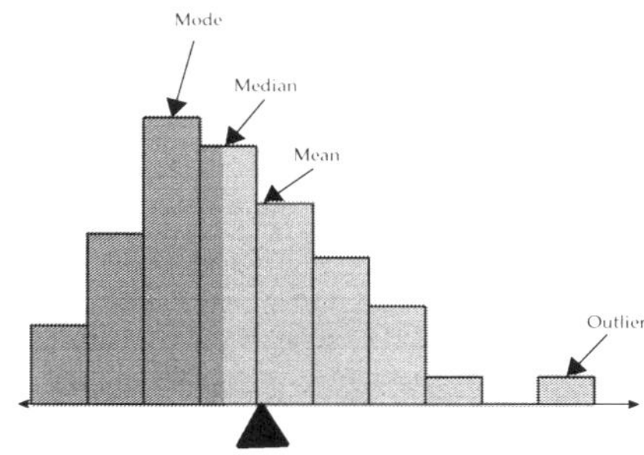

Part 1

For each pair of histograms, write a few sentences explaining why the one on the right is better than the one on the left.

a. Measured hemoglobin in 25 cats.

10.5	8.7	8.9	12	4.3
10	8.5	8	9.4	10.3
15.2	11	13	9.5	7
9.5	7.4	8.3	11.4	10.3
11.9	12.9	7.9	9.5	9.2

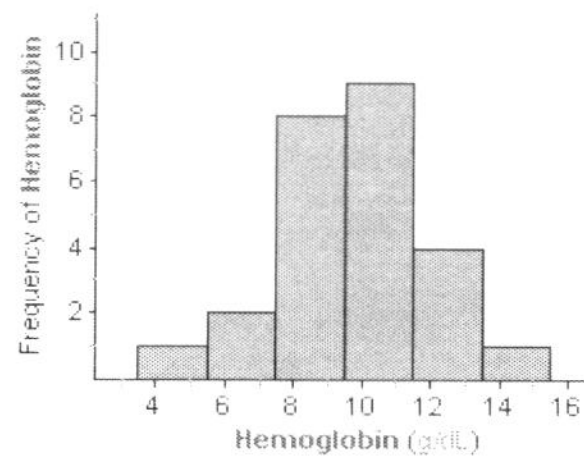

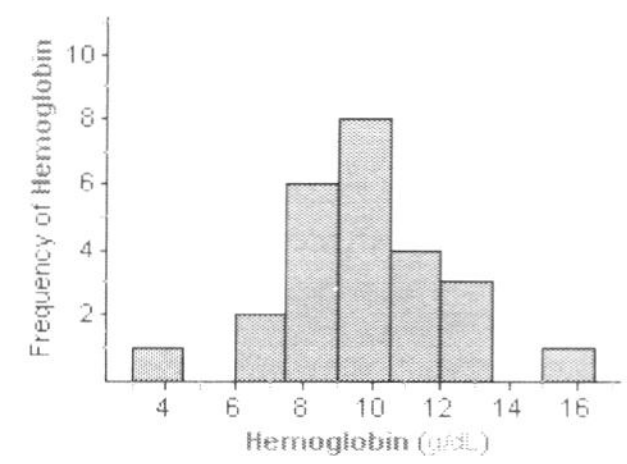

b. Hours worked last week by 25 community college students.

0	8.5	21	21.5	2
15.5	0	29	20.5	39
17	11	34	5.5	16.5
30.5	18.5	1	22.5	39
23	24	0	30.5	25.5

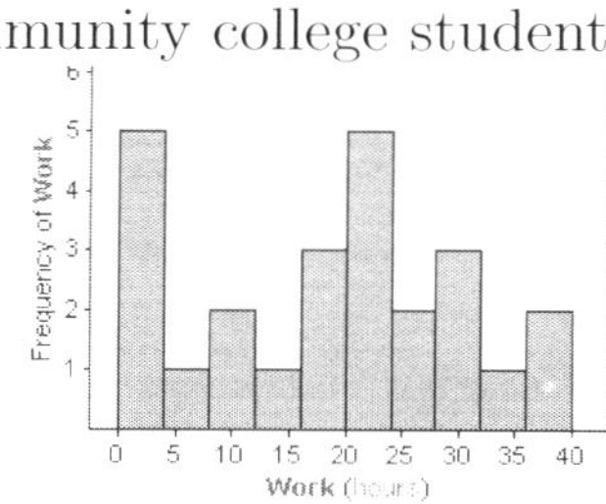

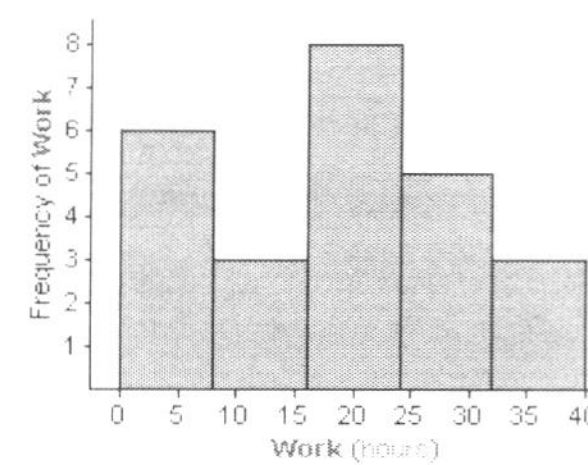

c. Sum of the roll of two dice (25 times).

10	8	8	12	4
10	8	6	9	8
2	11	3	9	7
9	7	4	7	6
5	5	7	6	7

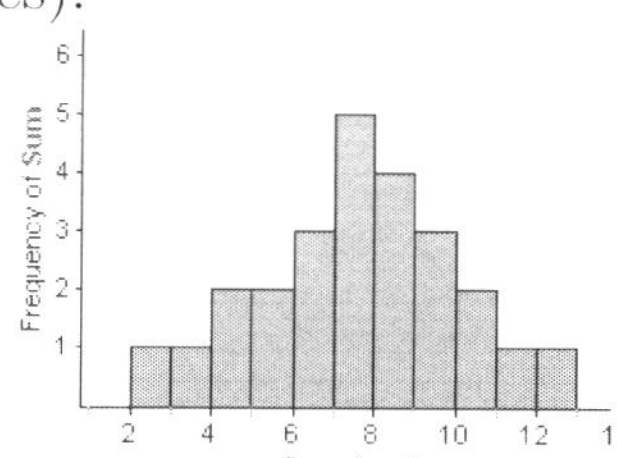

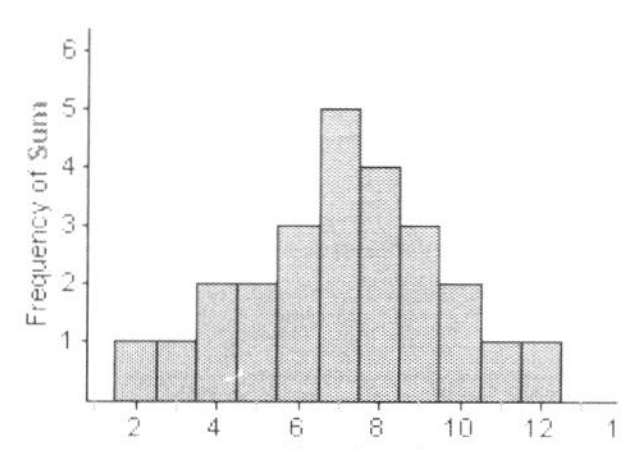

Part 2

For each histogram, write a few sentences explaining what is wrong with the graph and how to fix the problem.

a. Highway mileage for 30 compact cars.

28	26	30	24	30	28
32	26	34	21	21	26
30	25	29	30	33	26
22	36	28	26	34	32
35	35	38	34	39	33

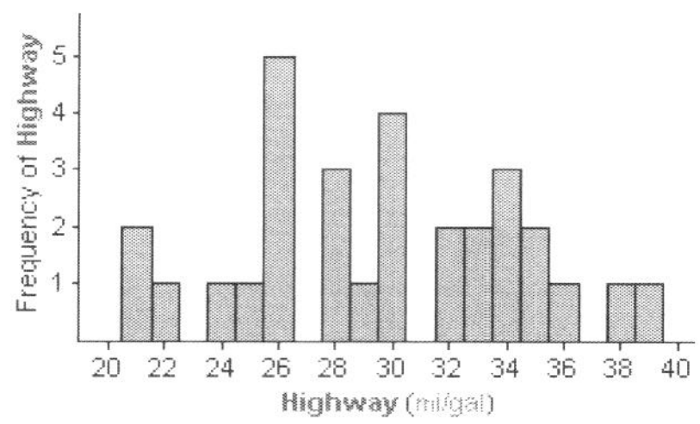

b. Time to complete a one hour math test (in minutes) for 30 students.

52	58	30	42	54	58
32	55	51	21	25	47
57	44	39	60	50	47
22	36	55	60	58	56
35	59	38	57	39	57

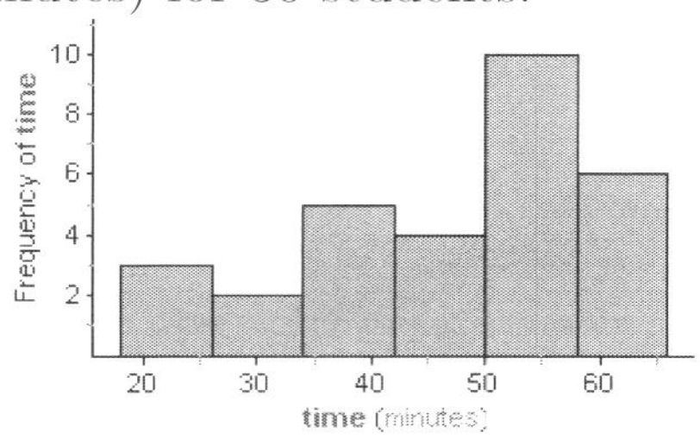

c. Cholesterol level of 30 young adults.

52	58	28	42	54	58
66	55	51	72	69	47
57	44	39	60	50	47
70	36	55	45	47	56
32	53	38	57	63	57

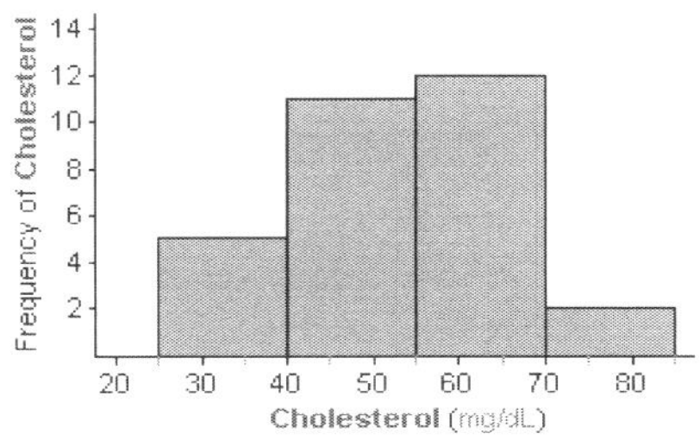

Part 3

Create a histogram for each data set. Explain the reasons for your choices of bin width and starting value.

a. The wait time (in minutes) for 60 randomly selected riders of the Demon Roller Coaster.

11	5	12	47	34	3	7	5	34	32	1	34	26	41	1
23	1	36	8	11	7	25	4	26	4	17	66	20	32	61
15	3	1	7	14	21	16	4	10	2	97	0	48	10	17
57	69	3	15	41	19	27	31	121	11	13	23	1	17	14

b. Amount of student loans (in dollars) taken for the 2014-15 school year for 20 college undergraduates.

2500	1000	2000	14000	1800	3800	10100	2200	900	1600
500	2200	6200	9100	2800	2500	1400	1320	750	12000

Teacher Notes for Bin There, Done That

Prerequisite Knowledge

- How to construct a histogram
- How to interpret a histogram

Objectives

The objective of this lab is for students to learn that not all histograms are equal. A second objective is for students to develop some familiarity with the histogram options in their graphing technology. Groups of 3 or 4 students should generate the conversations and ideas needed for this lab. If done completely out of class then allow 5 to 7 days to complete the lab. If you give them even part of one hour in class with you as a guide, you can shorten this lab to 3 days.

Comments and Answers

Part 1

a The histogram on the right shows the presence of potential outliers. These are hidden in the other histogram.

b The histogram on the left has too many bins. The shape of the distribution is not evident because of this.

c The position of the bins in the histogram on the right makes it clear that rolls that sum to 4 are in the bin with the 4 written below it.

Part 2

a This histogram has too many bins. Since the data values only go from 21 to 39, it might be better to have a bin width of 3 and to have bins centered on integer values.

b The bin start of this histogram should be 20.5. (As the max value is 60 the last bin should not go to 66.)

c This histogram should have more bins. Using a bin width of 5, starting on 25 would show the skewness of the data more clearly.

Part 3

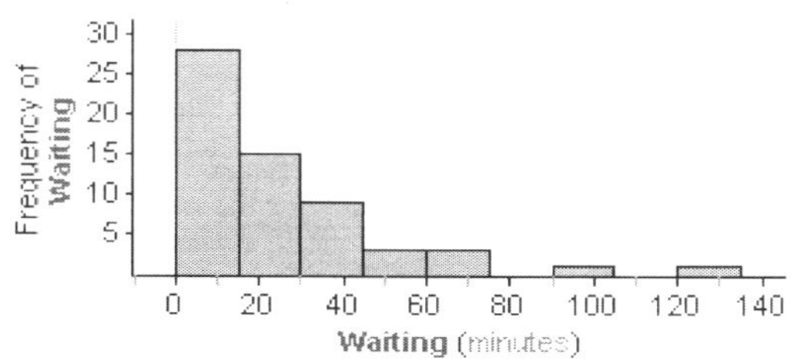

Students should choose a bin width that creates about 8 to 10 bins. The shape of the distribution should be clearly evident. The sample answer uses a bin width of 15 and starts at 0.

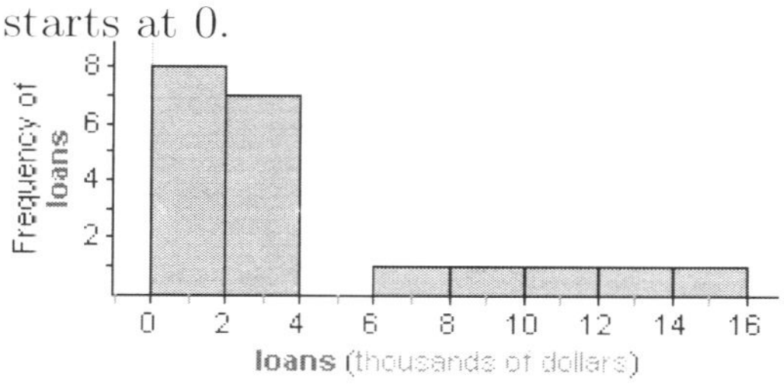

Students should choose a bin width that shows the nature of this distribution, indicating that most loans are smaller, and that the larger loans are for more varied amounts. No choices seem perfect in this case. To see the full range you must loose the shape of the distribution. This is a case where two histograms would help.

Lab 3
Finding Your Center

Suppose your instructor informs you on the first day of class that your performance will be evaluated this semester by ...

- Homework (100 points total)
- Weekly Quizzes (100 points total)
- Small group labs (100 points total)
- Test #1 (100 points total)
- Test #2 (100 points total)

What your instructor has not decided, is how to compute your final grade. This decision is up to the class to decide but the whole class must choose only one of ...

- The mean of the 5 scores
- The median of the 5 scores
- The mode of the 5 scores (When there is no mode you must average the two closest scores, when there are two modes you must average them)
- The 20% trimmed mean (also called the modified mean)
- The weighted mean (using 15, 15, 20, 25, 25 as the weights respected to the list above)

Part 1

For any set of 5 values some of these options are better than others.

a. Create a set of 5 values where the mean is the best choice of the 5. Calculate the final score using all five methods.

b. Create a set of 5 values where the median is the best choice of the 5. Calculate the final score using all five methods.

c. Create a set of 5 values where the mode is the best choice of the 5. Calculate the final score using all five methods.

d. Create a set of 5 values where the trimmed mean is the best choice of the 5. Calculate the final score using all five methods.

e. Create a set of 5 values where the weighted mean is the best choice of the 5. Calculate the final score using all five methods.

f. Describe the characteristics of sets of 5 values where the mean gives the highest value of the five choices. This is harder that it may first seem. Any five values that fit your description MUST have a mean greater than or equal to any of the other four options.

g. As in the last step, describe the characteristics of sets of 5 values where this time the median gives the highest value of the five choices. Any five values that fit your description MUST have a median greater than or equal to any of the other four options.

h. Now describe the characteristics of sets of 5 values where the mode gives the highest value of the five choices. Any five values that fit your description MUST have a mode greater than or equal to any of the other four options.

i. Describe the characteristics of sets of 5 values where the trimmed mean gives the highest value of the five choices. Any five values that fit your description MUST have a trimmed mean greater than or equal to any of the other four options.

j. Finally describe the characteristics of sets of 5 values where the weighted mean gives the highest value of the five choices. Any five values that fit your description MUST have a weighted mean greater than or equal to any of the other four options.

Part 2

Your group must choose which of the five you think is best and prepare an argument to convince the rest of the class that not only is your choice best, but also point out flaws in the four remaining choices.

Teacher Notes for Finding Your Center

Objectives

The objective of this lab is to understand measures of center, and how skew and outliers affect the different measures. There is no prerequisite knowledge for this lab. This lab works well for groups of 2 or 3 students. It can be done within a class period taking about 30 minutes. Or it can be assigned outside of class over 2 days.

Comments and Answers

Part 1

Sample solutions for a - e are found below.

f. The mean is highest when three or four scores are similar and the remaining one or two scores (not test scores) are much higher.

g. The median is highest when the top three scores are close but the remaining two are much lower and equal (or have the least distance between).

h. The mode is highest when the top two or three scores are equal (or have the least distance between) and the remaining three or two low scores are more spread.

i. The trimmed mean is highest when you have two high scores, two equal (or very close) sort of high scores, and one very low score.

j. The weighted mean is highest when three or four scores are similar and the remaining one or two test scores are much higher.

Part 2

We have seen good arguments for any of the five centers. Students will think about their personal history of grades when they make their choice. The flaw with the mean is that when there are one or two much lower values, the mean is always one of the lowest centers. (Skewed left data will move the mean lower.) The flaw with the median is the reverse. When there are one or two much higher values the median will fall below any of the means. The flaw with the mode is in a sample of 5 it is almost random as to where the two closest values will fall. This may make some students happy but probably make a few students feel they were badly cheated by this center. The flaw with the trimmed mean is occurs when there is one very high and/or two very low scores. The flaw with this weighted mean is for those students who don't test well, their average is now lower.

	HW	Quiz	Labs	Test1	Test2	Mean	Median	Mode	Trimmed	Weighted
a	100	80	80	80	80	**84**	80	80	80	83
b	100	95	90	70	70	85	**90**	70	85	82.25
c	100	100	85	70	70	85	85	**100**	85	82
d	100	99	90	90	60	87.8	90	90	**93**	85.35
e	80	80	80	80	100	84	80	80	80	**85**

Lab 4
Spread It On

Just like measures of center, there are several commonly used measures of variation, or spread, used to assess and compare data. But among statisticians there is only one that they really care about. As you are learning the subject this can be confusing because while there is only one value, there seem to be different formulas for it, and it may appear as if there is more than one of these measures.
The variance is the measure that we use, but it is difficult to explain or interpret to other people. It has units like square minutes or square square miles, that are rather meaningless. Thus, when we want to explain or interpret the value we take the square root and call this new value the standard deviation. It has the same units as the data and is therefore easier to interpret. However, both the variance and standard deviation tell us the same thing about the data set, so try not to think of them as separate measures. It is similar to when one person says 12 inches and another says 1 foot. You know they are both measuring the same thing.

Part 1

What effect does adding or subtracting a given value from each member of a data set have on the variance? This is the question you will answer in this part of the lab.

a. Create a set of values. Your set may be any size as long as there is more than one value in the set.

b. Use your technology to find the variance and the standard deviation of your set. (Your technology may offer you more than one choice for these measures and at this point it does not matter which one you pick. Just use the same one each time.)

c. Add 10 to every value in your set and find the two measures again. What do you observe?

d. Find the mean of your set and subtract that mean from every value. Again calculate the two measures. What do you observe?

Part 2

In physics the first moment is called the center of mass and it is the balance point of an object. In statistics the first moment is called the mean and it is the balance point of the data. In physics the second moment is called the rotational inertia of an object. A solid metal ball will spin just fine, but if we take the same metal and make it into a thin ring then it spins better and longer. The second moment is a value that expresses this property of the object. The second moment in statistics is called the variance and it is calculated the same way. To simplify the calculations we will center our data at zero. We can do this because as you found in Part 1, changing the center does not change the spread.

a. Consider all of the integers from -50 to 50. What is the mean (center) of these numbers?

b. Use your technology to square the integers from -50 to 50 and then find the average of the squares. This is called the variance or second moment. More correctly it is called the population variance.

c. Use the random number generator on your technology to select 10 random integers between -50 and 50. Then find the squares of the ten values and average these squares. How does this compare to your answer from b?

d. Working with your group, repeat this last step 100 times recording all 100 variances. How does the average of these 100 variances compare to the true variance you found in step b?

In this case we know the actual or true mean (called mu or μ) is zero. But in real life when we have a sample of data, we don't know the true mean. So when we calculate the variance from a sample we must use our estimate for the true mean, that is, our sample mean (called x-bar or $\bar{x}$).

e. Use the random number generator on your technology to find 10 random integers between -50 and 50. Find the squares of the ten values and average these squares.

f. Now find the mean of the ten values, subtract that mean from each original integer, then average the squares of these new numbers. How does this compare to value you got in the previous step?

g. Working with your group, repeat this last step 100 times recording all 100 variances. How does the average of these 100 variances compare to the true variance?

Your value from step g was probably smaller than the true variance you found in b. With a set of 10 data values, it was probably only about 90% of the true variance. It is as if we were averaging only 9 values instead of 10. This always happens. The value you calculate with a sample is smaller than the true variance. For this reason when you calculate the standard deviation from a SAMPLE, you divide by one less than the number of measures in your data set.

h. Write a short explanation of why a set of values would be closer to the mean of that set than to the mean of the entire population of values.

i. Since the set of values is closer to the mean of that set, they have a smaller spread from that mean than they do from the true mean. Thus, the value of the variance calculated using these values and their mean, is smaller than the true variance. How does dividing by one less than the number of measures in the data set compensate for this difference?

Part 3

a. Using a sample of 10 values determine if your technology uses the variance formula $\sigma^2 = \frac{1}{n}\sum_{i=1}^{n}(x_i - \mu)^2$ or $s^2 = \frac{1}{n-1}\sum_{i=1}^{n}(x_i - \bar{x})^2$. Some technology options will provide both. You should then identify which formula goes with which command.

b. Do the same for the standard deviation commands. (These are the same formulas as above only you take the square root when you are done.)

Recall that there is only one variance but the formula used depends on if the true mean is known or estimated.

c. Decide which is the better question to ask when you are asked to calculate a variance "Which formula should I use?", or "Is this a sample or is it the whole population?". Explain your choice.

Teacher Notes for Spread It On

Prerequisite Knowledge

- Finding the mean of a set of values
- Creating a list of random integers with technology

Objectives

The objectives of this lab are to understand the difference between variance and standard deviation, the difference between the formulas for population variance and sample variance, and develop some appreciation that these two are calculating the same thing. This lab can be done with groups from size 2 to 4. There is a lot of calculator function work but it goes quickly. About 3 or 4 days should give students time to complete the work and write the report.

Comments and Answers

Part 1

The value of the variance and standard deviation does not change when the data is translated.

Part 2

a. The mean is 0.

b. The average of the squares is 850.

c. The average will vary greatly depending on the integers selected.

d. By the time students have collected 100 sets of data values, the average of the variances will approach 850, or the true variance.

e. The average will vary here again but expect about 850

f. This value is smaller than that found in step e. It can be shown with calculus that the sum of squares is minimized when you subtract the sample mean.

We want to minimize $\sum_{i=1}^{n}(x_i - a)^2$ with respect to a.

Taking the derivative,

$\frac{d}{da}\sum_{i=1}^{n}(x_i - a)^2 = -2\sum_{i=1}^{n}(x_i - a)$.

This derivative is zero when

$\sum_{i=1}^{n}(x_i - a) = \sum_{i=1}^{n} x_i - \sum_{i=1}^{n} a = \sum_{i=1}^{n} x_i - na = 0$.

This implies $\sum_{i=1}^{n} x_i = na$ or

$a = \frac{1}{n}\sum_{i=1}^{n} x_i = \bar{x}$.

The second derivative, $\frac{d^2}{da^2} = 2n > 0$, so the sum is a minimum at $a = \bar{x}$.

g. By the time students have collected 100 sets of data values, the average of the variances will approach 765, or about 90% of the actual value.

h. One way to understand this is to think of selecting ten random numbers like throwing ten darts at a wall with no target on it. Then after throwing all of the darts, drawing a target with the bullseye at the center of your throws. Your darts will be closer to this center than to any other bullseye you can put on the wall. That is, your values will be closer to their own mean than any other, including the population mean.

i. Dividing by a smaller number makes the result larger.

Part 3

a.

Technology	s^2	σ^2
TI-83/84	variance()	
TI-89	variance()	varPop()
TI nspire	varSamp()	varPop()
Excel	VAR()	VARP()
Fathom	variance()	popVariance()

b.

	s	σ
TI-83/84	stdDev()	σ
TI-89	stDev()	stDevPop()
TI nspire	stDevSamp()	stDevPop()
Excel	STDEV()	STDEVP()
Fathom	s()	popStdDev()

c. It is probably better to ask if it is a sample or an entire population. If you know it is a population, then you can apply the population formula that you know. Likewise with a sample. If you ask which formula to use, you may get a version of a formula that you are not familiar with. Thus you still may not know what to do.

Depending on your textbook or resource you can find these three forms for the the formulas for the sample variance.

$$s^2 = \frac{1}{n-1}\sum_{i=1}^{n}(x_i - \bar{x})^2$$

$$s^2 = \frac{1}{n-1}\left(\sum_{i=1}^{n} x_i^2 - n\bar{x}^2\right)$$

$$s^2 = \frac{1}{n-1}\left(\sum_{i=1}^{n} x_i^2 - \frac{1}{n}\left(\sum_{i=1}^{n} x_i\right)^2\right)$$

They are all equal, yet some texts prefer one form over the other. So asking what formula can lead to some real confusion.

Lab 5
Thinking Outside the Boxplot

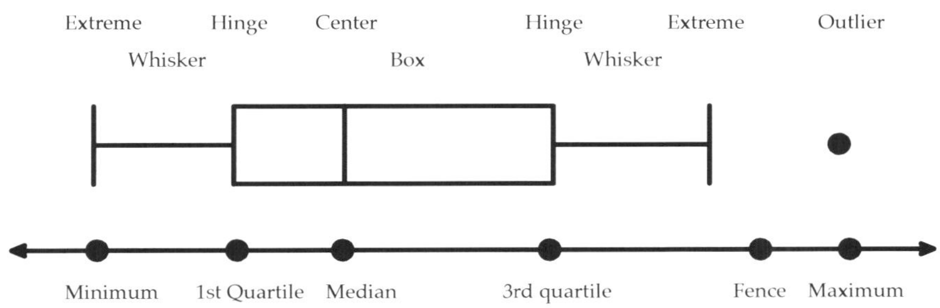

Box and Whisker plots have existed since the 1970's. At that time technology was scarce and difficult to use. The plot was developed by John Tukey, because it was simple to construct by hand from a set of ordered data. It is one of the most popular plots to use when comparing multiple distributions or samples. But it is not a perfect plot. The chief flaws of this plot are...

- It does not show the sample size.
- It does not clearly distinguish if the distribution is mounded, flat, or bimodal.
- It shows potential outliers, but may hide real outliers, or make non-outliers seem like they are outliers.
- It uses medians and quartiles while most statistical analysis uses means and standard deviations.

Your challenge is to design a "better" box plot that shows more information than the current one, and to explain your design.

Part 1

a. Create several sample data sets and make standard box plots for them. Be sure to use sets of different sizes and with different types of distributions.

b. Create a new type of box plot that shows more information than the standard one. Look at the flaws listed above and choose one or two to try to overcome.

Part 2

a. Write a careful, complete description of how to create your new box plot. Use at least one of your sample data sets to help you explain the steps.

b. Explain why your new plot is better than the standard one.

Teacher Notes for Thinking Outside the Boxplot

Prerequisite Knowledge

- Measure of center and spread
- How to construct a boxplot
- Concepts of skew and outliers

Objectives

The objective of this lab is to have students think about the structure of the box plot and become familiar with the vocabulary associated with the plot. Actually, designing a new plot is just a tool to achieve this. Groups of size 3 or 4 work well here. This lab should take 5 to 7 days.

Comments and Answers

Part 1

The sample data sets should include sets with both odd and even sample sizes. They should include a variety of shapes of distributions. The box plots for these should be clearly labeled and correct. The new box plot should include the same type of information as the original. That is, it should show a measure of center, spread, give an indication of outliers, and skew. Here is a sample solution using the data sets
A: 1, 1.5, 2, 2.5, 3, 4, 5, 7, 9
B: 1, 2, 2.5, 3, 3, 3, 3.5, 3.5, 4, 5, 6, 6.5, 6.5, 7, 7, 7, 8, 9

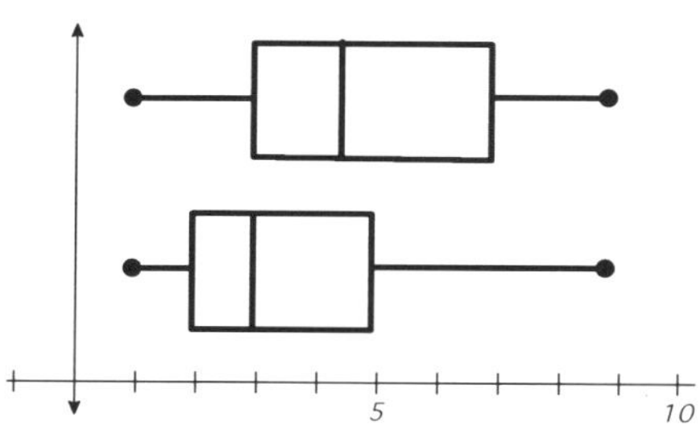

The previous plots are the standard boxplots for the samples, the improved plots are below.

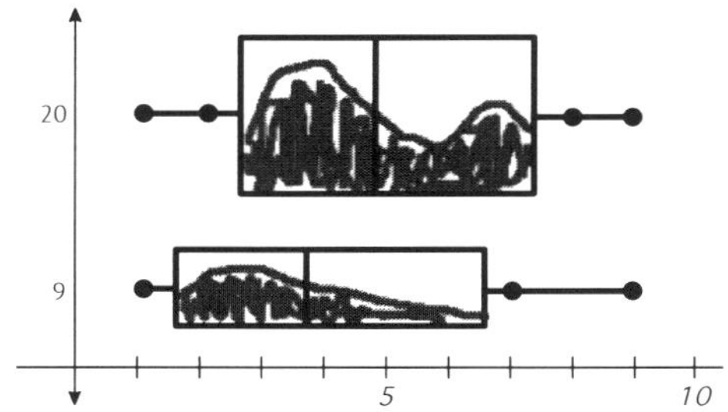

- The height of the box and axis label show the sample size.
- Shading inside the box indicates mound(s) and denser areas of data
- Points showing all tail values are added to the whiskers.
- The center line is the mean and the hinges are one standard deviation from the mean.

Part 2

The explanation of how to create the box plot should be clear and easy to follow. One technique to determine this, is to give a sample data set and these instructions to another group of students and see if they can replicate the plot. The description should correctly use the vocabulary throughout. The explanation of the superiority of the new plot should reference at least one of the flaws mentioned in the beginning of the lab. You might consider having each group present their improved box plot procedure to the class and then taking a vote on which one the class thinks is best. As students prepare to do this, it will strengthen their understanding of the concepts involved and also give them experience in trying to use the vocabulary clearly and concisely.

Lab 6
Getting the Drop on Association

Equipment needed

- Rubber bands (one package)
- Water bottle
- Meter stick or tape measure

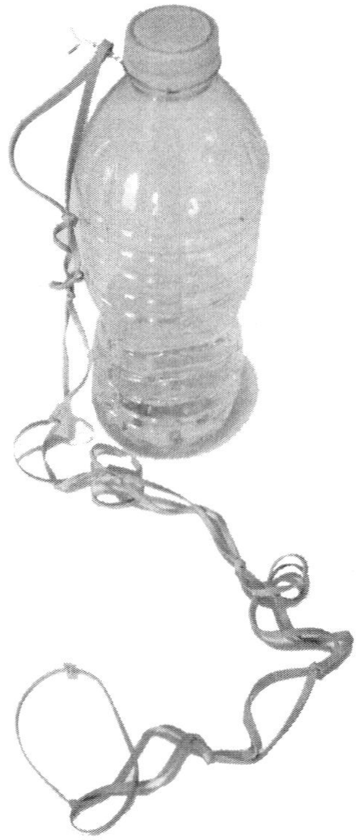

Part 1

a. Choose four values at random between 2 and 12. Build rubber band chains of these four lengths. These are mini-bungee jumping cords.

b. Now you need a weight such as a half-filled water bottle, to serve as your subject. Attach the weight to your first chain and drop the weight from the point where you are holding the other end of the chain. Record the maximum distance the cord stretches. That is, from holding point to the bottom of the water bottle.

c. Repeat this with each of the other three chains you made. This set of four ordered pairs $(ChainLength, MaxDrop)$ is your data set for this lab.

d. Make a scatter plot of your data.

e. Describe in words the strength of the association between these two variables, chain length and maximum drop.

f. Using the four values for chain length, find the mean and standard deviation ($\bar{x}$ and s_x).

g. Using the four values for maximum stretch, find the mean and standard deviation ($\bar{y}$ and s_y). These means and standard deviations will be used in Part 2.

Part 2

Like measures of center or spread, there is more than one measure of association between two numeric variables. In this lab you will look at four of these measures. Two of them are model dependent. That is, they depend on the type of function used to describe the relationship. The other two are linear only measures. That is, they are only meaningful when the relationship between the variables is a linear relationship. If we think our data from Part 1 is linear, then all four will work. We will discuss what to do if the data is not linear in Part 3.

a. You may recall learning the formula for the variance when you learned about measures of spread.

$$s_x^2 = \frac{1}{n-1}\sum_{i=1}^{n}(x_i - \bar{x})^2 \qquad \text{and} \qquad s_y^2 = \frac{1}{n-1}\sum_{i=1}^{n}(y_i - \bar{y})^2$$

A similar measure is called the covariance.

$$s_{xy} = \frac{1}{n-1}\sum_{i=1}^{n}(x_i - \bar{x})(y_i - \bar{y})$$

Plot the four points (1, 1); (1, 3); (5, 1); and (5, 3). Then show the steps to evaluate the covariance formula using the four points.

b. When calculating the variance all terms of the sum are positive because of the squaring. But covariance has both positive and negative terms. Consider which points contributed positive terms and which contributed negative terms to the covariance. How are these two sets different? What does it mean about the points if the sum is positive, negative, or zero?

c. Calculate the covariance of the ordered pairs $(ChainLength, MaxDrop)$ from Part 1. What does this tell you about your data?

d. Plot these four collinear points (1, 4); (3, 8); (7, 16); (9, 20) and find the covariance. Also find the standard deviation for both the set of x-coordinates and the set of y-coordinates. Then find the product of these two standard deviations.

e. The covariance, as a number, is hard to interpret and generally is only used when comparing different sets of data (i.e. which has the larger covariance and how much larger it is). But the covariance is never larger than the product of the two standard deviations, and it only will equal that when the data is perfectly linear. Our second measure of association is a scaled version of this covariance.

$$r = \frac{s_{xy}}{s_x s_y}$$

What are the largest and smallest this value can be?

f. This measure is commonly called the correlation. Find the correlation of the data from Part 1, from step a of Part 2, and from step d of Part 2. What do these three values tell you about those three sets of data?

g. For the model-dependent measures of associations, once you have a model, you can address the strength of association by how close the data lies to the model. The word "close" implies some measure of distance. In statistics, where the measures are linked in a control / response relationship, the distance is typically between the measured response and the response predicted by the model. Graphically we place the response variable on the y-axis and measure the error, or residual, vertically, as the observed minus the predicted for a particular control value. In a data set with strong association the errors are all near zero. As the errors can be both positive and negative (above or below the model) explain why it is not useful to use the sum of these errors to measure the association.

h. The third measure of association is the variance of these residuals, called the mean square error, MSE, or its companion the root mean square error, RMSE.

$$s_e^2 = \frac{1}{n-2}\sum_{i=1}^{n}(y_i - \hat{y}_i)^2 \qquad \text{and} \qquad s_e = \sqrt{s_e^2}$$

Use the values from the data in Part 1 and find the least squares model ...

$$\hat{y}_i = \bar{y} + r\frac{s_y}{s_x}(x_i - \bar{x})$$

Record your equation and your four values for $\hat{y}$. Now calculate the four residuals.

i. Calculate the MSE and RMSE. Explain what this value tells you about the association of these data.

j. This last measure of association gets smaller when the association is stronger. As with the covariance, the MSE can be hard to interpret. The largest value that this will have is when there is no association. In this case the errors will be the same as the deviations of the response (y-values). In order to create a measure that gets larger when the association is stronger we use

$$R^2 = 1 - \frac{(n-2)s_e^2}{(n-1)s_y^2}$$

This value is called determination. Find the determination for the data from Part 1.

k. When the data is linear and the model is the least squares line, then $R^2 = r^2$, that is, the determination is equal to the square of the correlation. Is that true for your calculations?

Part 3

These last two measures, MSE and determination, can be found after any type of model is selected for the data (polynomial, exponential, even multi-variable). But the first two measures, covariance and correlation, are meant only for linear data. When you have non-linear data you can still use these measures but you have to create an alternate data set based on your values first. This process is called linearization. In essence, we "straighten" a curve using functions and inverse functions on the data until we have a new set of data that can be described as linear. Here is a very simple example of this.

a. Suppose you thought that instead of using the equation $\hat{y} = a + bx$ to describe your bungee stretch, you thought that $\hat{y} = a + bx^2$ is more likely the model. If this is true then a plot of $(ChainLength, MaxDrop)$ will look like a curve. But a plot of $(ChainLength^2, MaxDrop)$ will appear linear. Make a new plot of these four points.

b. This new set of data is now the linearization of your data. Use these values to calculate the covariance.

c. Now calculate the correlation of the these new data. What does this tell you about this set of data?

d. What does that mean to you about your choice of model when you compare this to the correlation you found in Part 2?

This was a very simple type of linearization. Usually this part is a bit more complicated but the idea is that we can use covariance and correlation along with linearization to help us select models that are valuable.

Teacher Notes for Getting the Drop on Association

Prerequisite Knowledge

- Scatter plots and least squares lines.

Equipment needed

- Rubber bands
- Water bottle
- Meter stick or tape measure

Objectives

The objective of this lab to understand how the measures of association are found and related to each other. This lab works with 3 or 4 students in a group. You can give up part of a day (about 20 minutes) to collect the data in Part 1. If this lab is assigned while the class is learning about correlation and other measures of association, then make the lab due when you finish the unit. If it follows the unit, then give 4 or 5 days to complete the lab.

Comments and Answers

Part 1

The maximum number of bands will depend on the size and type of bands. You don't want the drop to be longer than the height of the room if you are trying to collect data in the classroom. To attach the bottle to the bands we have used short piece of wire or a twist-tie on the neck of the bottle. Sample data: (10, 236); (4, 112); (7, 168); (6, 148). This sample shows strong linear association.

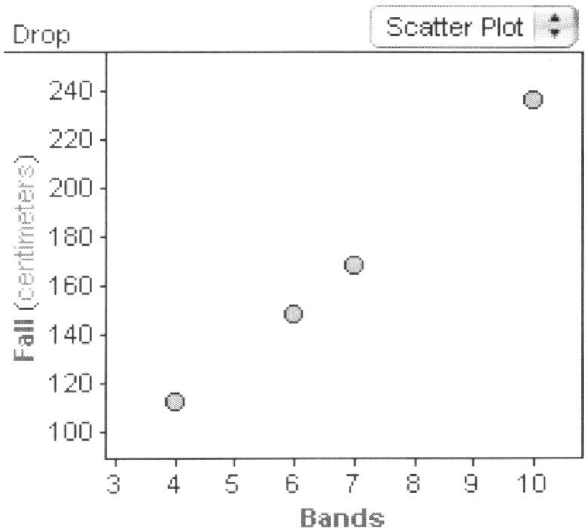

	Bands	Drop
$\bar{x}$	6.75	166 cm
s	2.5	52.1 cm

Part 2

a. $\bar{x} = 3$ and $\bar{y} = 2$ so
$s_{xy} = \frac{1}{3}(2 + -2 + -2 + 2) = 0$

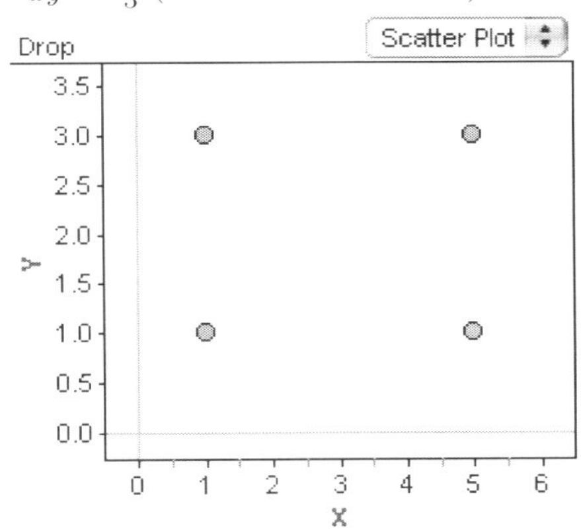

b. The lower left and upper right are positive. Connecting these points is a line of positive slope. The other two points connect with a negative slope. When the sum is positive then the trend in the data has a positive slope. When the sum is zero there is no trend to the data.

c. For the sample data, $s_{xy} = 130$. This

tells us that the trend in the data is a positive slope.

d. $s_{xy} = 26.6667, s_x = 3.65148,$ $s_y = 7.30297, s_x s_y = 26.6667$

e. The value of r, the correlation, will be -1 when negatively sloped points are collinear and +1 when positively sloped points are collinear.

f. From the sample data, $r = 130/130.256 = 0.9980$. For step a, $r = 0$ and for step d, $r = 1$. There is a strong association in the data that was collected (but not quite perfect).

g. The sum of errors will be zero or near zero like the sum of deviations.

h. The least squares model for the sample data is

$$\hat{y}_i = 166 + 0.9980 \frac{52.1}{2.5}(x_i - 6.75)$$

$$\hat{y}_i = 25.6 + 20.8x_i$$

Predicted points $(10, 233.6)(4, 108.8)(7, 171.2)(6, 150.4)$. Errors 2.4, 3.2, -3.2, -2.4

i. The mean square error is

$$s_e^2 = \frac{1}{2}(2.4^2 + 3.2^2 + 2.4^2 + 3.2^2) = 16$$

The root mean square error is 4. This is to say, predictions lie within 4 cm of the data measured. The data must be associated strongly to have such a close mean prediction.

j. The determination for the sample data is

$$R^2 = 1 - \frac{(2)16}{(3)2714.6667} = 0.99607$$

.

k. The square of correlation should be equal to determination if students did all of the calculations correctly.

Part 3

a. This plot was made with sample data after the chain length had been squared.

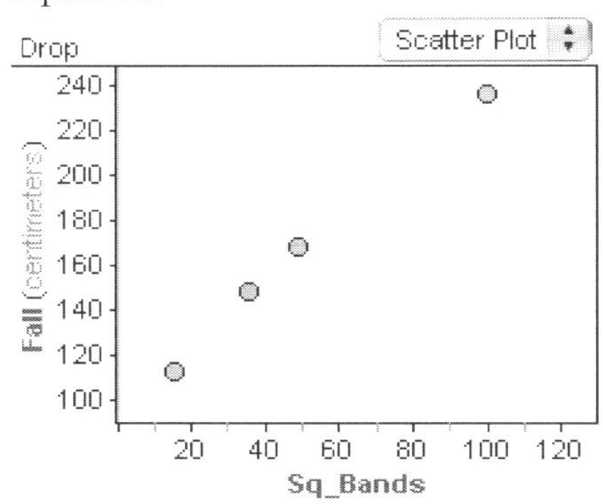

b. The covariance of the sample data is 1862 cm.

c. The correlation of the sample data is 0.9972. There is a strong association between these two variables.

d. The correlation was stronger for the original data that for the linearization of the quadratic data. So a linear model is a better model than the quadratic model.

Lab 7
Throw Me a Curve, Bell

Probabilities and calculations based on a normal distribution are very important in a first course of statistics. This often involves much formula work, calculations, and technology. With all this, a misplaced decimal or dropped negative sign can make a error leading to a completely wrong conclusion. Maintaining a visual image of what you are doing will help you avoid this type of mistake. Even experienced statisticians will make a sketch of what they are solving to verify that the answer makes sense.

Equipment needed

- Wire (under 30 cm in length)
- Ruler (with mm markings)

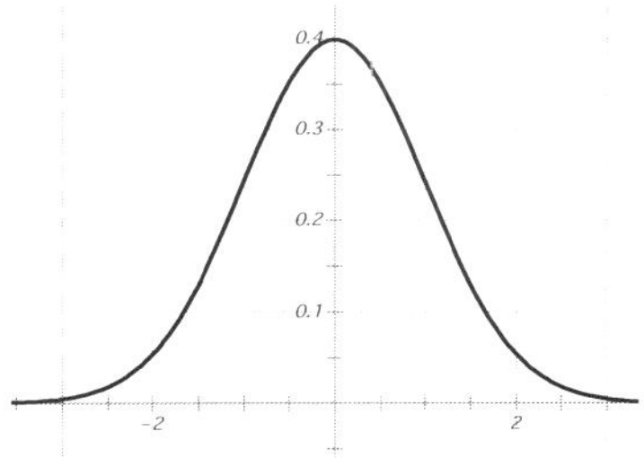

Part 1

a. You will collect data from about thirty persons. Each person will measure the length of the same piece of wire using the same ruler. Have each person record the length to the nearest 0.01cm (0.1 mm). The ruler is only marked in mm so, they will have to estimate the last digit. Tell the measurers that the last digit is required so they should do the best they can.

b. Prepare one or two distribution plots of the data. Comment on the possible "normality" of this data set.

c. Find the sample mean and standard deviation of the measurements.

Part 2

It is often assumed that measurement errors follow a normal distribution. In this part you shall determine if this is valid with the measurements you collected.
Often a crude sketch of a normal curve is sufficient. Even an inverted parabola can often serve as a normal curve graph. But in this lab you are going to spend time and energy drawing a normal curve accurately, even if you never need do so again. Here are the steps to sketching an accurate normal curve:

a. Start with a horizontal axis and mark the mean value in the center of the axis.

b. On the axis mark and label values that are one, two and three standard deviations from the mean to both the left and right of the mean. Your axis should now have 7 equally spaced values on it.

c. The height of the curve is not critical to most estimations coming from the graph, though in an accurate graph, the height is dependent on the standard deviation. Because the area under the curve is fixed, an increase in the standard deviation has the inverse effect on the height of the curve. When the standard deviation is one, the maximum height is about 0.4. If the standard deviation is s then the height is 0.4 divided by s. Using your standard deviation, mark a point at this height above the mean.

d. At one standard deviation, right and left of the mean, the height is 60% as tall or $\frac{0.24}{s}$. Mark these two points on the graph.

e. At two standard deviations, right and left of the mean, the height is 12.5% as tall or $\frac{0.05}{s}$. Mark these two points on the graph.

f. At three standard deviations, right and left of the mean, the height is 1% as tall or practically touching the axis. Mark these two points on the graph.

g. Connect these last seven points making a smooth bell curve.

Part 3

a. The three numbers associated with normal distributions are 68%, 95%, and 99.7%. These numbers form the empirical rule. Using the numbers below your graph in Part 2, write three complete sentences about a normal distribution of data.

b. Compare the data you collected with the statements you just wrote. Comment on how the data relates to these statements.

c. The seven marks on the horizontal axis divide the area under your curve into 6 regions. Draw vertical segments from each of these marks to the curve. Use the empirical rule in (a) to find the approximate area of each region and add this information to your graph.

d. Using only your graph, estimate the area that is more than 1.5 standard deviations above the mean. Explain your reasoning.

e. Using only your graph, estimate the area that is below 0.5 standard deviations below the mean. Explain your reasoning.

f. Using only your graph, estimate the area that is between 2.5 standard deviations below the mean and the mean. Explain your reasoning.

Part 4

Multiple choice questions are well written when a person who understands the concepts can quickly identify the correct answer and each of the foils (wrong answers) is the result of some basic misunderstanding. In this part you will write a multiple choice question that ...

a. Concerns probability and the normal distribution.

b. Needs only estimation and a carefully drawn graph to answer correctly.

c. Needs no normal CDF calculations from technology to select the right answer.

d. Has the other three foils coming from some simple misunderstanding.

When you have written the question and four answers then ...

e. Identify the right answer and explain how to find it.

f. For each of the three foils explain how a person might come up with that particular answer.

Teacher Notes for Throw Me a Curve, Bell

Prerequisite Knowledge

- Continuous distribution graphs
- The empirical rule

Equipment needed

- Wire (under 30 cm in length)
- Ruler (with mm markings)

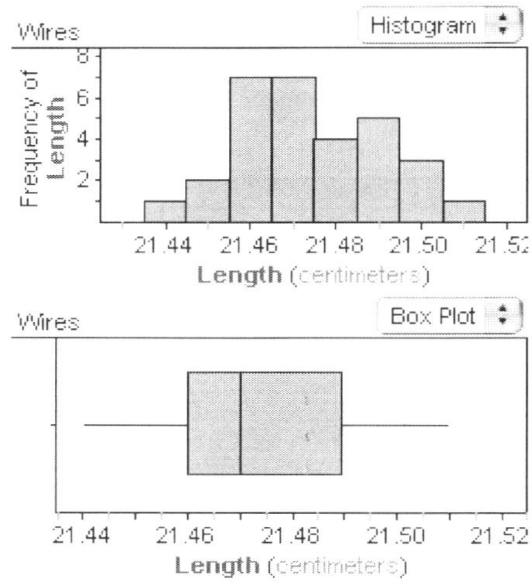

$\bar{x} = 21.4743, s = 0.0171572$ cm

Objectives

The objective of this lab is to help students become more familiar with the normal distribution and probabilities associated with it. If they are collecting their own data then use groups of 3 or 4, If they are provided with sample data then groups of 2 or 3 will work. Students should have 5 to 7 days to complete the lab, with 2 fewer days if they are given data. Understanding how multiple choice questions are written can help students be more efficient in solving them.

Comments and Answers

Part 1

This data should look somewhat normal though it depends on the ability and seriousness of the people making the measurement. If you are saving time you may provide students with sample data such as that found at the end of these notes. The answers provided here are done using that sample data as examples of what to expect.

Part 2

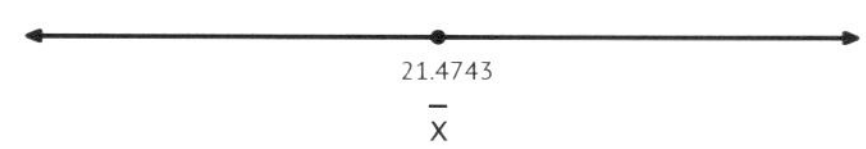

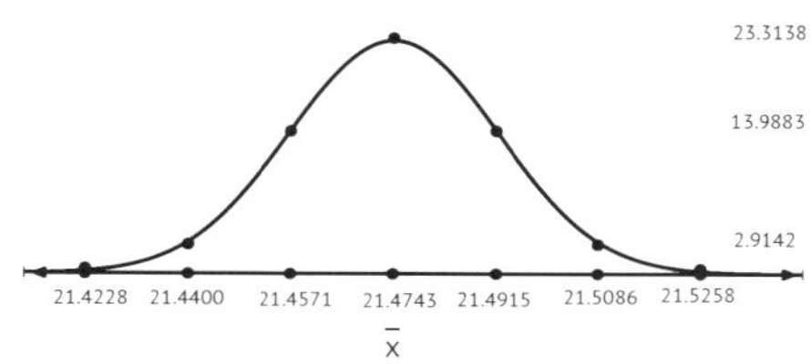

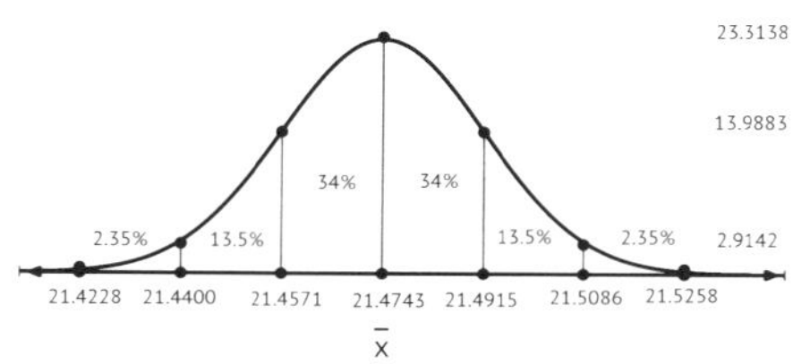

Part 3

You would expect 68% of the measures between 21.4571 cm and 21.4915 cm
You would expect 95% of the measures between 21.4400 cm and 21.5086 cm
You would expect 99.7% of the measures between 21.4288 cm and 21.5258 cm

The data show 23 of 30 or 76.7% are within one SD of the mean.
The data show 29 of 30 or 96.7% are within two SD of the mean.
The data show 30 of 30 or 100% are within three SD of the mean.
These are reasonable for a sample from a normal distribution.

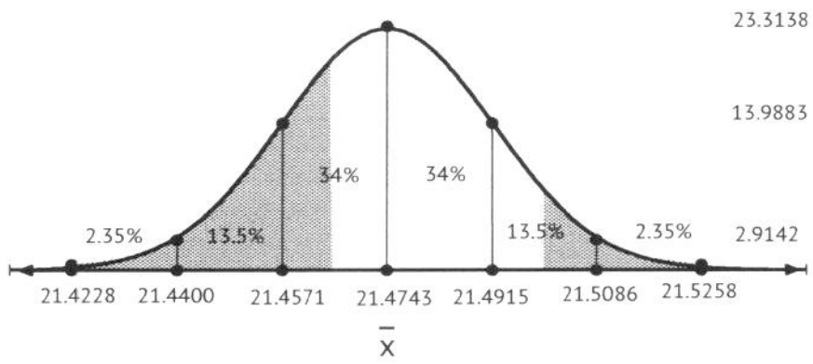

The area greater than 1.5 SD above the mean is between 2.35% and 15.85%. They might estimate this about 7 or 8%.
The area less then 0.5 SD below the mean is between 15.85% and 50%. They might estimate this about 25 to 30%.
The area between 2.5 SD below the mean to the mean is between 47.5% and 50%. They might estimate this about 48 or 49%.
An interesting twist on this is to view areas as "near" triangles. Using the area of a triangle with the length of base and height gives an estimate of the shaded region. In general this is not a practical or very accurate way to estimate area, but it can help some students make connections to what they see on the graph.

Part 4

You might consider using one or two of the best questions on the next test. Announcing this may encourage students to work harder on creating quality questions.
Most questions written will likely be estimations of area or probability because that is what they worked on here. You may also see the reverse where given an area, or probability, they are asked to estimate a value or interval. Most common foils are to go the wrong direction calculating one minus the area, or to start on the wrong side of the mean and go in the correct direction. There is also the one sided question with the two sided answer or its reverse. You may also see problems where the deviation is treated like a Z score while forgetting to divide by the SD.

Table 7.1: Sample Data: Lengths in centimeters

21.5	21.49	21.47	21.47	21.49	21.46	21.45	21.46	21.47	21.46
21.48	21.46	21.51	21.49	21.49	21.46	21.48	21.47	21.46	21.47
21.46	21.5	21.47	21.48	21.44	21.49	21.48	21.47	21.45	21.5

Lab 8
The Apple and the Tree

Have you heard the expression "The apple does not fall far from the tree."? We can estimate the probability of a random event if we know the distribution function of that event. Consider the event of how far from the tree an apple falls. This is a continuous variable, so the probability distribution function (pdf) is a continuous function. In this lab we try to match the probability function of this event and calculate some probabilities.

Equipment needed

- A handful of small disks or pennies
- A clear area of floor
- A tape measure or meter stick

Part 1

The size of the area you need depends on your disk and floor type. If you have a carpeted floor, then a circle of 1 meter radius is usually enough area. If you have a hard surface, then you may need a 3 meter circle.

a. Place a marker on the floor. Hold the disks in your hand 1 to 2 meters directly above the marker. Release the disks and quickly get out of the way.

b. Measure the distance from the marker to the nearest edge of each disk. If a disk is covering the marker then that measure is zero. Retrieve each disk after it has been measured.

c. Record all your measures and enter these values into your technology. Sort them from smallest to largest.

d. Calculate the mean and standard deviation of this data set, then make one or two distribution plots of your data. Comment on your distribution plots.

Part 2

The goal is to create a graph that compares our data to a known distribution (One that you guess is right). This works pretty well for DISproving bad guesses, and sort of well for "proving" good guesses. To see this, begin with a bad guess: a normal distribution. Next you will generate a theoretical, or perfect, data set based on that distribution.

a. Your "perfect" set will be the same size as your sample data, n. Next you must calculate the percentile values for each data value in the set. If you divide 100% by n you determine how far apart all your percentiles are. For example if $n = 25$ then each data value will be 4%ile higher than the previous one in an ordered list. The first data value will be at a percentile equal to half of the difference between values. For instance, if $n = 25$, the difference is 4%, and your first data value has a percentile rank of 0.02 and the next 0.06. Create a list of percentile values for each position in your perfect set of data values. Express each percentile as a decimal between 0 and 1.

b. Using the inverse of the normal cdf function (with the estimated mean and standard deviation) find the value corresponding to each percentile in your list. This is your perfect data set. In other words, these are the theoretical positions for each disk IF the distribution were normal.

c. If the distribution of your data is normal, then the positions you calculated in b SHOULD be close to your data measurements. Make a graph of data values vs. theoretical values and add the line $y = x$ to the graph. Comment on what you have found and what it means to you.

Part 3

Now you are going to repeat this procedure with a better guess, an exponential distribution. To estimate this distribution you only need one parameter, the mean. Now you will generate a theoretical, or perfect, data set based on that distribution.

a. Begin with the same percentile list you found in Part 2 since that has not changed.

b. Use the inverse of the exponential cdf function, with the estimated mean, to find the value corresponding to each percentile in your list (If your technology does not have a built in inverse exponential distribution function then use the formula $x = \mu \ln(1 - p)$ where p is the percentile and μ is the mean.) These are the theoretical positions for each disk IF the distribution were exponential.

c. Again make a graph of data values vs. theoretical values and add the line $y = x$ to the graph. Comment on what you have found and what it means to you.

Part 4

The probability for an exponential distribution can be found $P(a < x < b) = e^{-a/\mu} - e^{-b/\mu}$

a. What is the probability that a disk will fall less then 10 cm from the center?

b. What is the probability that a disk will fall between 10 and 20 cm from the center?

c. What is the probability that a disk will fall more than 20 cm from the center?

d. What radius is likely to contain one half of the disks that fall?

How do these values compare to what you observed in your data? Comment about your comparisons.

Teacher Notes for The Apple and the Tree

Prerequisite Knowledge

- Continuous Distributions
- Normal Distribution
- Z-scores

Equipment needed

- Between 25 to 50 small disks or pennies
- A two meter diameter carpeted floor or 6 meter diameter hardwood floor
- A tape measure or meter sticks

Objectives

The objective of this lab is to have students construct both normal and exponential probability plots to evaluate the type of distribution. These plots are not the typical ones produced by software but serve the same purpose. Typical software plots match data against a "standard" normal or "standard" exponential distribution, rather than the specific distribution used in this lab. The advantage of using the specific distribution is that the values make more sense to students. Once they understand this plot they can transfer this understanding to the more abstract plots we normally see. This lab works well with groups of 3 or 4 students. It is possible to do the data collection as an in-class activity and then have all groups share the same data set.

Comments and Answers

Part 1

After dropping the disks, you should see most disks near the center and only a few disks at extreme distances. Data collection goes quickly if one student holds the tape or meter stick at zero, another reads the measure, a third records the data, and a fourth collects disks after they are measured. Make sure students sort these values or Part 2 will not work. These plots were made from the sample data found at the end of the notes.

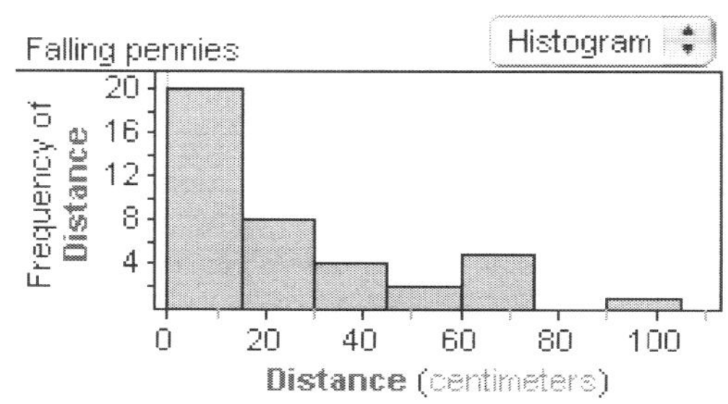

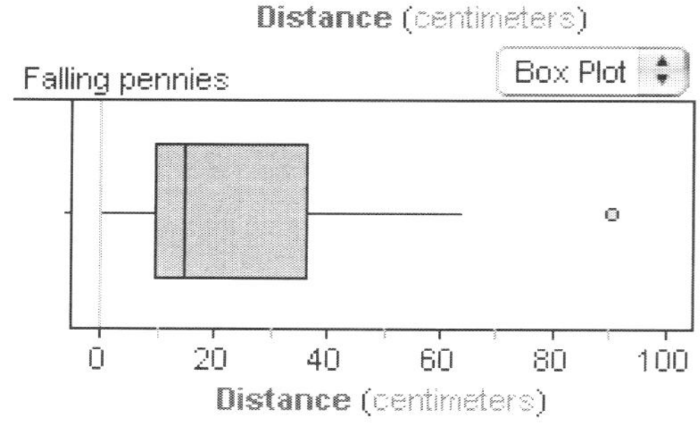

$\bar{x} = 24.3825, s = 22.3555$ cm

This should look like exponential data and not at all like normal data.

Part 2

Your students may wonder why they start at 1/2 the percentile width. As many distributions have no lower or upper bound both the 0%ile and 100%ile are often undefined. Thus the usual way is to split the difference and start a half width from 0% and end a half width from 100%.

a. This can be done quickly using the sequence functions in your technology. Enter the command, $\text{seq}((x-0.5)/N, x, 1, N) \to L_2$, where N is the number of data values, to fill list 2 with the percentiles.

b. This can be done one value at a time or done quickly using an inverse normal CDF function. On the TI84 this looks like $\text{InvNorm}(L_2, \mu, \sigma) \to L_3$. Where L_2 is the list of percentiles, μ is the mean, and σ is the standard deviation.

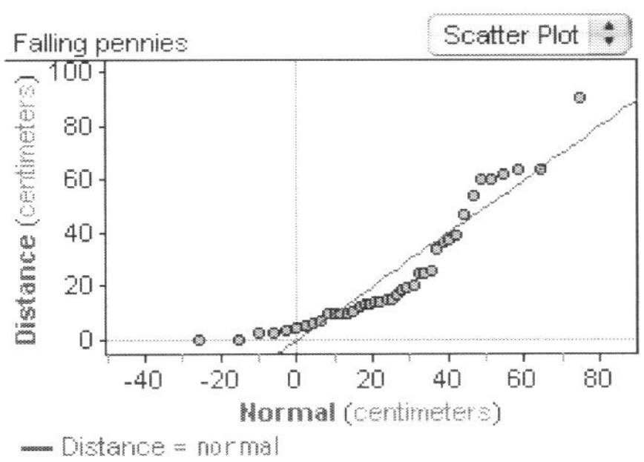

c. These points do not follow any line, much less the line $y = x$. It is clear that this data is not normal.

Part 3

a. See 2a

b. Again this can be done using a command like $-\mu \ln(1 - L_2) \to L_4$ to determine all of the estimated distances and place them in a list.

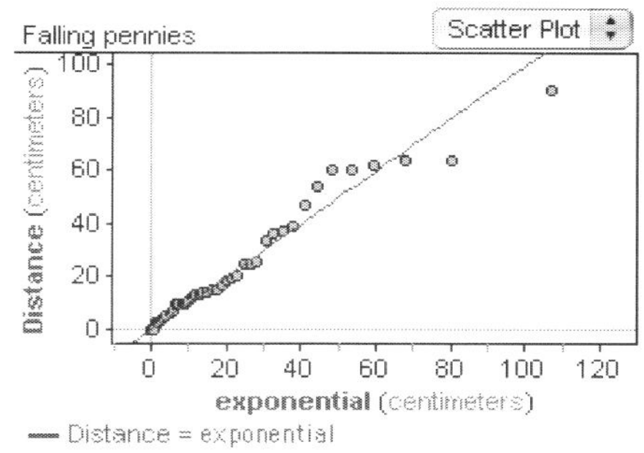

c. Typically the first 75% of the data will fit well and the last quarter of the data will be hit and miss. Hopefully they will recognize that this is as much as can be expected for this type of verification and accept the exponential as a reasonable pdf for this event.

Part 4

Here students must use the CDF of the exponential distribution to find the probability. These estimates are based on the mean of 24.38 from the sample data.

$$P(dist < 10) = 1 - e^{-10/\mu} \approx 0.336$$

$$P(10 < dist < 20) = e^{-10/\mu} - e^{-20/\mu} \approx 0.223$$

$$P(dist > 20) = e^{-20/\mu} \approx 0.440$$

$$P(dist > r) = e^{-r/\mu} = 0.5$$

$$-\mu \ln 0.5 \approx 16.9 \text{ cm}$$

The sample data has 13 values (or 32.5%) below 10 cm, twelve values (or 30%) between 10 and 20 cm, and 15 values (or 37.5%) above 20 cm.

Table 8.1: Sample Data: distances in centimeters

0	0	2.3	2.4	3.6	4	5.5	5.9	7.4	9.5	9.6	9.8	9.9	10.9
12.3	13.4	13.5	13.7	13.8	14.9	15	17.1	18.6	19	19.9	24.5	24.6	26
33.5	36.6	36.8	39.2	46.5	53.7	60.4	60.5	62.4	64	64	90.6		

Lab 9
Is This Normal or Not?

You have solved problems using facts and formulas about the normal distribution. You have learned the characteristics of this distribution and how to recognize it. But it can still be a tough call when looking at sample data to decide if a distribution is normal.

Equipment needed

- Stopwatch
- Video clip

Part 1

You will collect data from between 20 and 30 persons. If you are lucky, you might get them all to do it at once. But it doesn't matter, you can collect from several small groups or even one person at a time.

a. Find a short video clip. In this clip there must be a scene that lasts between 5 and 10 seconds. There should clear boundaries for when the scene starts and stops, with no fading in or out.

b. Have your subjects find the stopwatch/timer app on their phone. Have them practice starting and stopping the timer a few times.

c. Explain that they will be timing the length of the scene as accurately as they can. You may need to show them the scene once before timing, so they will know what they are looking for.

d. Now play the clip as they time the scene. Have them record the time with as much accuracy as their device gives them.

Part 2

Using graphs, statistics, and attributes of your sample, prepare an argument to convince a reader that this sample likely comes from a normal distribution. Find at least 3 features of your set that help convince someone these times are from a normal distribution.

Part 3

Using graphs, statistics, and attributes of your sample, prepare an argument to convince a reader that this sample is not likely to come from a normal distribution. Find at least 3 features of your set that help convince someone these times are not from a normal distribution.

Teacher Notes for Is This Normal or Not?

Prerequisite Knowledge

- Attributes of Normal distributions

Equipment needed

- Stopwatch
- Video clip
- About 20 to 30 subjects

Objectives

The objective of this lab is for students to assemble all they know about the normal distribution. As the data collected will not be perfect, there will be room for discussion as to whether or not the data comes from a normal distribution. The task here should provide ample opportunity for discussion. If groups are puzzled, feel free to direct their attention to various features of the distribution to help them assemble their arguments. The take away from this lab is not to make them more confused but to show that a sample cannot be classified as normal or not normal. It can only show indicators of such a distribution. Most samples they will see will look normal enough, but that does not mean that they are or are not normal. It does mean that probabilities based on the normal distribution are probably OK, unless the probability is out in the tail of the distribution.
This can work for almost any size group. If data is collected in class, or sample data is used, this activity can be completed in less than one period. If assigned out of class, 3 to 5 days should be enough time to complete it.

Comments and Answers

Part 1

At the time of the printing of this book, there is a 18 second YouTube called "Math Joke". In this there is about a 6 second scene with a green-blue background. This was used to collect the sample data. If time is limited you may wish to use this sample data that is found at the end of the notes.

Part 2

The data collected will likely be fairly normal so it should not be difficult to find reasons to show normality. Here is a short list of common comments you may see.

a. The distribution is mounded.

b. The distribution is not skewed (or only slightly skewed).

c. The mean and median are close (mean = 6.154 sec, med = 6.13 sec).

d. The histogram is bell shaped.

e. Some part of the empirical rule is related (for the sample data they might say 74% of the data is within 1 SD of the sample mean, this is close to the 68% expected for a normal distribution.

f. A normal probability plot is fairly linear.

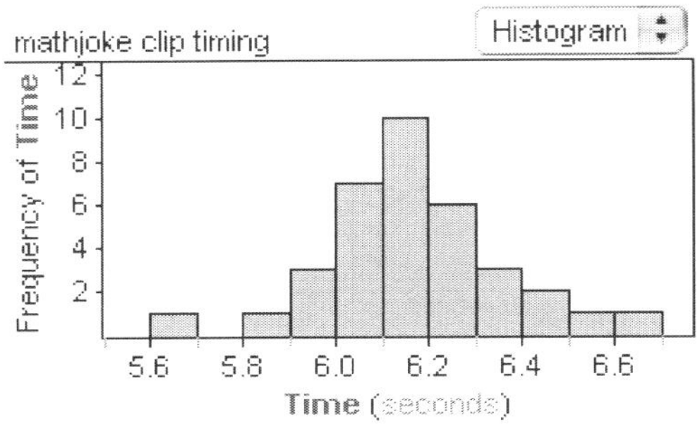

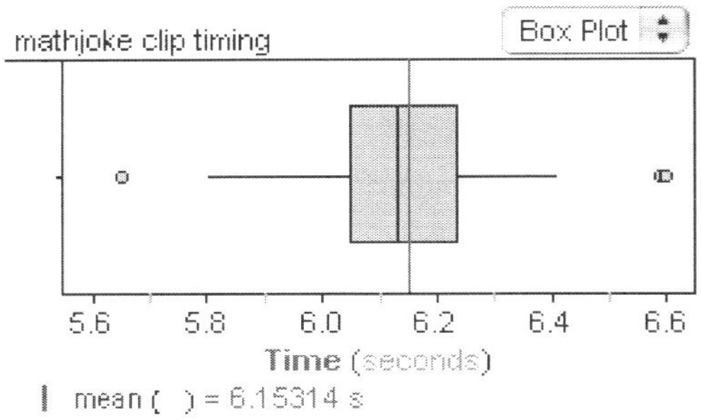

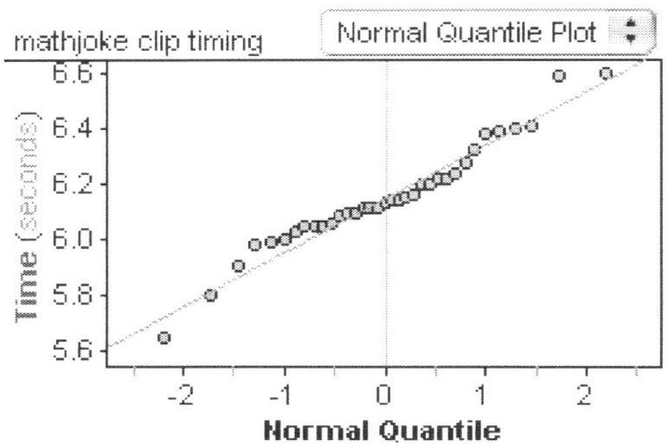

Part 3

The sample would be very unusual if there was nothing that was not "normal" about it, so this list should also not be hard to come up with once they put their mind to it. It is even reasonable that something might make both lists.

a. The distribution is slightly skewed.

b. The mean and median are not equal (mean = 6.154 sec, med = 6.13 sec).

c. The histogram is not a perfect bell shaped.

d. There are outliers.

e. Some part of the empirical rule is wrong (for the sample data they might say 91% of the data is within 2 SD of the sample mean, this is not close to the 95% expected for a normal distribution).

f. A normal probability plot is not exactly linear.

Table 9.1: Sample Data: times in seconds

5.65	5.8	5.91	5.98	5.99	6	6.03	6.05	6.05	6.05	6.06	6.09
6.1	6.1	6.11	6.11	6.11	6.13	6.14	6.14	6.15	6.16	6.2	6.2
6.22	6.22	6.24	6.28	6.32	6.38	6.39	6.4	6.41	6.59	6.6	

Lab 10
By Chance or By Choice

You are not as random as you think. Often experimenters, data collectors, even scientists will use the term "randomly selected" when in fact they have chosen subjects or values themselves, believing they have done so "randomly". In this lab we will put this to a test.

Part 1

a. Obtain a single sheet of lined paper and create two columns on the paper. Label the columns A and B.

b. Flip a coin. If the coin lands heads up, then you will first fill column A. If it lands tails up, then you first fill column B.

c. Without using any tool, document, or device, fill the thirty or so lines in your chosen column with integers between 1 and 100. Try to be as random as you can. Your goal is to fool someone into thinking that this column was created by random process.

d. Now use a computer, calculator, or random number table to produce random integers from 1 to 100, and copy these into the other column.

e. Number each paper from your group but make sure there is no indication of which column was created by the human and which came from technology. On a separate piece of paper identify each of the sheets and which column on each sheet that was done first. Submit all the papers from your group to your instructor.

Part 2

In this part you will study random numbers created by technology (or tables). Generate a dozen or so sets of random integers between 1 and 100, equal in length to the lists in Part 1. Study these lists and look for what might make them look different than a human made list. Comment on each of these items about your lists.

a. How often do you see the same value show up more than once in a list of random numbers? How many numbers are repeated?

b. How often do you find numbers that are close together (a difference of 5 or less), appearing consecutively in the list?

c. If you put the numbers in each list in order, how wide is the largest empty interval in each list?

d. How often do the random samples include the values of 1 or 100 or both?

e. How often is the second digit of a truly random number 0? or 5? or 7?

Part 3

Your instructor will provide your group with a number of sheets equal to your group size that were created by other group members. You may each take one sheet but you should still act as a group, discussing and swapping papers to check each other. You must investigate the two columns of numbers on each sheet and decide which one you believe to be the human created column. You will want use graphs as well as other kinds of analysis to decide which list is which.

Part 4

Your grade for this lab will be partly determined by how many lists you correctly identify as human created, and also by how you use statistical tools in your analysis in making your choices. Your report should include for each sheet that you were given ...

a. The column you think was human made.

b. The features of this column that made you think it was not from a random number source.

c. Support each argument with the graphs or statistical measures that helped in your decision.

Teacher Notes for By Chance or By Choice

Prerequisite Knowledge

- How to get random numbers from your technology or a random number table.

Objectives

The objective of this lab is to gain an understanding of what randomization is all about. Humans are all about patterns. We see patterns even when there are none. We make patterns even when we are trying not to. It is just how we are wired together. Statistical sampling DEPENDS on randomization so patterns in data collection are not good.
Groups of 3 or 4 work well for getting the discussion of ideas going in this study. We have typically done Part 1 in class, spending about 15 minutes on it. The rest of the lab can be done using class time or out of class. If using class time it will take nearly an hour and you will be very busy sorting and organizing while they are working on Part 2 so you won't be able to help them much. If done out of class give them 5 to 7 days assuming you get them the papers they need for Part 3 by the second or third day.

Comments and Answers

Part 1

When groups hand in their collected lists, be sure to add a group identifier to the number at the top of each paper. It will be important to know the identity of each human-created list. Reorganize the sheets so that each group gets back the same number of papers as they turned in. The papers they receive should not come from their own group, and it is best if they come from a mix of two or more groups.

Part 2

These sample answers are based on a list of 30 random values between 1 and 100.

a. There is less than 1% chance that no number is repeated. On average you should see about 4 values repeated and there is about a 25% chance that one value shows up 3 times in a list.

b. There is less than a 5% chance that consecutive values will never be within 5 of each other. For most lists you should expect to see this happen about three times.

c. It is rare to see a list without a gap of 8 within the distribution. The typical gap is about 14 with gaps up to about 18 to 20 not too uncommon.

d. A little less than half the lists will have either 1 or 100 in them, about one sixth of the lists will have both 1 and 100.

e. The second digit should have equal frequency for all, thus about 3 values should end in zero and three end in 5 and three end in seven. It is rare that you have NO values for a particular last digit and also rare that you have 6 or more values ending in the same digit.

Part 3

Encourage students to consider all the questions in Part 2 as they analyze lists of integers. It may not be one thing that makes it look human, but rather two or three "unlikely" occurrences that help you choose the human list.

Lab 11
Census Sensibility

You are now working for the US Census Bureau. The Bureau frequently sends out survey teams to gather information. You must provide the Bureau with careful plans on how to gather information from a small town. They will provide you with a map of the town and a listing, alphabetically by street, of every residence in the town. Almost all residences in this town are occupied. The list contains 3650 residences.
The census bureau will send 2 two-person team contracted to work three days for 10 hours each day (9 to 9 with 2 hours for meals). The survey takes an average of 15 minutes to complete for each residence and the Bureau hopes to have 200 completed surveys after 3 days.

Part 1

You will have very limited time once you receive the map and the residence list so you need to prepare a plan of what you will do once the information arrives. Be as detailed as you can in your plan as if you were giving instructions to the team collecting the data.

Part 2

The USCB has just contacted you to let you know that within the 3650 residences there are 285 small apartments. All of these apartments can be found on only three streets in this town. In the numbered listing of residences they show up in three intervals: 805-864, 933-1027, and 3110-3239.
Since the Bureau believes that these residences will be quite different in such things as average income, they have asked you to make sure that these locations are neither underrepresented or overrepresented in your survey. Repeat what you did in Part 1 ensuring that this new requirement is met.

Part 3

Another communication from the USCB tells you the first two groups of apartments are located in the center of town. Most of these are occupied by young persons in their 20s, while the last group includes two retirement complexes at the edge of town that are occupied by persons age 70 or older. Since lumping this data together would actually give you a poorer estimate for mean age, they now want you to treat them separately, still ensuring that each group is properly represented in your sample. Please provide a third plan that accomplishes this goal.

Teacher Notes for Census Sensibility

Prerequisite Knowledge

- Simple random sampling
- Stratified random sampling

Objectives

The objective of this lab is to understand random sampling, stratified random sampling, and how to communicate a sampling plan clearly. Each of these topics can be challenging for students. If you make all three parts (plans) due at the same time, it is likely that either all three will be good or all three will be poor. If you have groups turn them in one at a time, you can return each one with comments. Then most groups will have a well explained and complete plan by part 3. You can also make it a class activity to critique Part 1 and 2 plans by covering the names and displaying or distributing the plans for class analysis. Groups of 2 to 4 can work on this. If there are 3 in a group, then you can alternate who writes up each part. We would do this over a full week. If, as recommended, you collect the Parts one at a time, give two days for the first Part to be due. If you can't read, mark, and return these in a day then at least spend a few minutes of class time talking about what was missing or wrong with the plans. Now give two more days before Part 2 is due. Repeat with two or three days for Part 3.

Comments and Answers

Part 1

This is a simple random sample (SRS). Be sure the plans are clear in how they will use randomization to select residences to survey. They must select 200 values from the 3650. Since some of the locations may be vacant, the groups should select more that 200 so that the 200 quota can be reached. Once the households are selected they should be mapped so that the interview teams will have an organized route through the city. In that way little time is wasted crossing town to the next residence. Planning should be made for the teams to double back and revisit locations that have no answer on the fist visit.

Part 2

This is a stratified random sample. As 7.8% of the residences in the town are apartments, there needs to be an SRS of 16 (plus two or three) apartments selected and an SRS of 184 (plus 5 or 6) non-apartments selected. This is easiest to do this is by picking a number from 1 to 285 then mapping that value to the apartment in the intervals of the first list. You can create a table/formula that tells you which number matches which residence. (i.e. if x is less than 60 add 804, if less than 155 add 873, otherwise add 2955). Or you can do this as a single SRS of the population, then go through the list until you have reached both the quota for apartments and non-apartments by hi-lighting the apartments. As before these should be mapped and routed efficiently. The survey team should be told to stop visiting apartments when they have 16 even if there are more apartments on the list. The route planned should make sure that the apartments are not at the end of the planned route in case they don't go as quickly as planned.

Part 3

This is also a stratified random sample. Now 4.2% are in-town apartments, 3.6% are retirement apartments, and 92.2% are not apartments. So you need an SRS of 8 in-town apartments (plus 2 for empty apartments), an SRS of 7 (plus two) retirement apartments, and an SRS of 184 (plus 5 or 6) non-apartments selected. As before these should be mapped and routed efficiently. The survey team should be told to stop visiting apartments when they have 8 in-town and 7 retirement apartments even if there are more apartments on the list. The route planned should make sure that the apartments are not at the end of the planned route in case they don't go as quickly as planned.

Lab 12
The Inch Pinch

In this activity you will perform the same experiment in two ways and compare the results and accuracy of these results.

Equipment needed

- 30 cm (12 in) ruler
- about 12 cm of tape

To build your tool you need to fold the tape and attach it to the 0 cm end of the ruler so the "T" is about 4 cm wide.

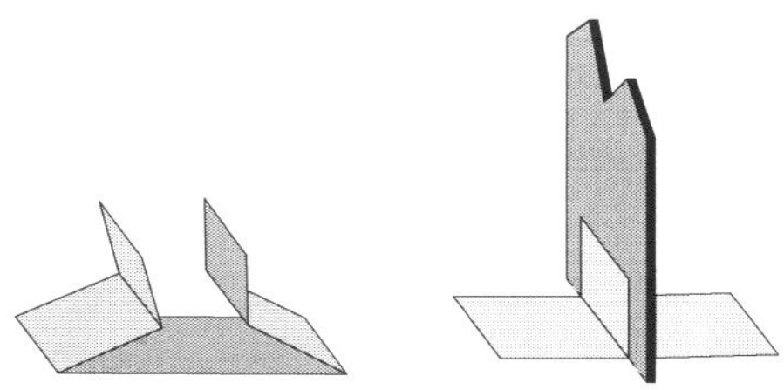

Part 1: Independent Sampling

The first method requires 20 subjects. Each subject is to catch a dropped ruler using either their dominant hand or their non-dominant hand.

a. How will you assign which hand each subject should be using?

b. Once assigned, ask the subject to hold their thumb and first finger just touching the edges of the tape as you hold the ruler. Instruct them to pinch the ruler as soon as it is released. After a random number of seconds, release the ruler and then record the ruler marking just above their thumb.

c. Repeat this for each subject. Retain the same roles for your group members, having the same person do all of the dropping and the same person make all the measurements in order to keep things as consistent as possible. (Don't forget to record which hand each subject uses.)

d. Calculate the mean for all 10 subjects in each of your two groups. Then calculate the difference in the means; non-dominant mean minus dominant mean.

e. Calculate the variance for each of your sample sets. Consider these two groups to be independent and sum the variances. Use the square root of this sum to find the standard deviation of differences.

Part 2: Matched Pair Sampling

The second method requires 10 subjects. Each subject is to catch a dropped ruler twice; once using their dominant hand and once using their non-dominant hand.

a. How will you assign which hand each subject should use first?

b. Once assigned, ask the subject to hold their thumb and first finger just touching the edges of the tape as you hold the ruler. Instruct them to pinch the ruler as soon as it is released. After a random number of seconds release the ruler and then record the ruler marking just above their thumb.

c. Repeat this for the other hand of the same subject. Retain the same roles for your group members, having the same person do all of the dropping and the same person make all of the measurements in order to keep things as consistent as possible.

d. Repeat this process for each subject. (Don't forget to record which hand they use for each drop.)

e. Calculate the difference; non-dominant minus dominant for each of the 10 subjects. Then calculate the mean of the differences for all 10 subjects.
$E(N - D) = E(N) - E(D)$.

f. Calculate the variance of your 10 differences and compute the standard deviation of the differences. $V(N - D) = V(N) - V(D)$.

Part 3: Comparison

In each experiment you obtained a mean difference and a standard deviation of the differences. In your report comment on the following points:

a. Which experiment gave "better" results, and what made the results better?

b. Since one of these is better, why not use that method in all data collection?

c. Give an example where you would have to use the "poorer" method.

d. When looking at two columns of data it is not always possible to tell which of the methods was used. When you read the description of the data collection, what are the indicators that tell you which method?

e. What does "randomize" mean for matched pair sampling, and why would it matter if you did not randomize?

f. What does randomize mean for independent sampling, and why would it matter if you did not randomize?

Teacher Notes for The Inch Pinch

Prerequisite Knowledge

- Ability to find means and variances

Equipment needed

- 30 cm (12 in) ruler
- about 12 cm of tape

Objectives

One objective of this lab is to learn to distinguish between independent sampling and matched pair sampling. Another objective is to understand the difference in randomization of the two methods. A final objective is to recognize the superiority of the matched pair sampling and when it can be used.
This lab works well with groups of 3 or 4 students and takes about 7 to 10 days to complete, unless you provide students with sample data.

Comments and Answers

Part 1

For randomization, students typically use a coin flip, such as assigning heads for dominant hand and tails for non. The problem with this is that you can finish one set of ten early and then the last three or four are the same hand. One solution is the coin flip is used for persons 1, 3, 5, 7 ... then persons 2, 4, 6, ... are the other choice. Randomizing the time before releasing the ruler is also important, as there may be a relationship between the time to pinch the ruler and how long it is before the ruler is released. Thus you should encourage students to not rely on their own ability to release the ruler at random times, but use some method of determining how long to hold the ruler for each drop. These results are based on the sample data found at the end of these notes.
$\bar{x}_N = 17.165$ cm, $\bar{x}_D = 15.614$ cm
$s_N^2 = 2.909$ sq cm, $s_D^2 = 2.591$ sq cm,
$\bar{x}_{dif} = 17.165 - 15.614 = 1.551$ cm
$s_{dif}^2 = 2.909 + 2.691 = 5.600$ sq cm
$s_{dif} = \sqrt{5.6} = 2.366$ cm

Part 2

In this Part a coin flip is the easy and best answer to assigning which hand should go first.
Using the sample data the differences are 1.69, 2.72, 0.89, 1.38, -0.41, 2.03, 1.59, 2.24, 1.04, 1.10 cm
$\bar{x}_{dif} = 1.427$ cm
$s_{dif}^2 = 0.746$ sq cm
$s_{dif} = \sqrt{0.746} = 0.864$ cm

Part 3

a. While the two means are likely about the same, we find the spread of the independent samples is far greater than that of the matched pair differences. In the sample data the standard deviation is more that 2.5 times as large. This indicates that our mean of the differences is more standard deviations from zero than the difference of the means. For the sample data, the mean of the differences is $\frac{1.427}{0.864} = 1.652$ standard deviations from zero, while the difference of the means is $\frac{1.551}{2.366} = 0.656$ standard deviations from zero.

b. It is not always possible to collect two data values from the same subject, so we can't always do a matched pair design.

c. Answers will vary here. They may observe that collecting data may destroy the thing you are measuring as in measuring the force needed to break something, or the time it takes to cure an ailment.

d. Note that if the two columns are of different lengths, it is going to be independent sampling, but if they are of equal sample size, you must look to see how many subjects were used to determine if they measured one value for each subject, or two values for each subject.

e. In matched pair sampling, randomizing is done to vary the order that you get your two measurements. If you always took measure A before measure B, then you cannot rule out the possibility that some "learning" happened. One does get better at the ruler drop game with practice. So the difference between the measures could reflect the learning, not the true difference in the hands.

f. In independent sampling, randomizing is done to make sure that assignments are random and that measuring is also random. If you let people decide which hand they will use, then people with little non-dominant control would likely choose their dominant hand while those who work well with both hands would choose the non-dominant and the data would be biased. If you tested all of one group first, then if you become better or different in the way you measure at some point in the process, this would show up more in the second group than the first.

Physics

An object will accelerate when dropped, so the distance will likely not have a normal distribution as the distance is related to the square of the time. If you wish to convert distance to time use $d = \frac{1}{2}gt^2$ with $g = 980$ cm/sec^2 or 384 in/sec^2 thus $t = \sqrt{\frac{d}{490}}$ sec when d is in centimeters or $t = \sqrt{\frac{d}{192}}$ sec when d is in inches.

Sample Data: All measurements are in centimeters.

Independent Non Dominant	Independent Dominant	Matched Non Dominant	Matched Dominant
17.63	13.08	14.86	13.17
18.71	17.72	20.54	17.82
17.95	16.04	15.44	14.55
16.07	16.06	16.52	15.14
19.50	13.19	18.22	18.63
14.89	15.03	23.54	21.51
14.46	14.87	17.87	16.28
17.64	15.90	15.53	13.29
18.72	17.70	21.64	20.60
16.08	16.55	16.16	15.06

Lab 13
A Rolling Die Gathers No Data

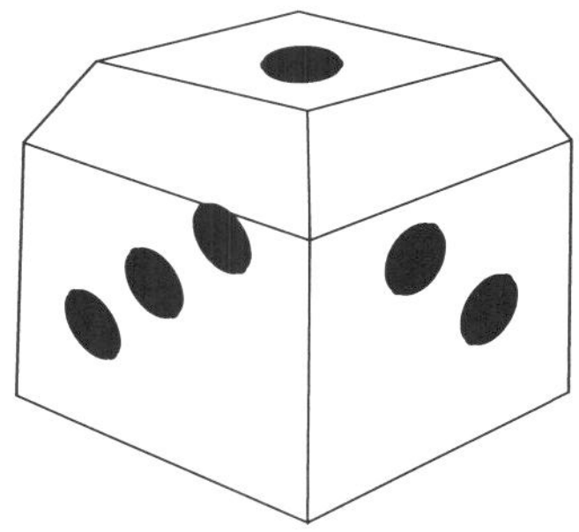

Equipment needed

- Foam Die
- Box cutter or exacto knife

Part 1

a. Trim the four edges around the one pip face of the foam die. Try to trim each edge in the exact same way. Do not trim any other edges.

b. Roll this die 100 times and record the number of pips on the top face for each roll.

c. Use these experimental results to calculate estimates of the probability for each possible roll. Create a probability distribution chart for the outcomes of this experiment.

X	1	2	3	4	5	6
$P(X)$						

d. Use the formula for the expected value of a discrete distribution to calculate the mean value for a roll of this die.

e. In the same way calculate the variance and standard deviation of the distribution.

Part 2

a. Use the values from your distribution in Part 1 to calculate the probabilities for all 36 possible outcomes for rolling two identical dice.

Y	X	1	2	3	4	5	6
X	$P(X)$						
1							
2							
3							
4							
5							
6							

b. Based on the numbers from your table, calculate an estimate of the probability for each possible sum. Create a probability distribution chart for the sum of two trimmed dice.

Y	2	3	4	5	6	7	8	9	10	11	12
$P(Y)$											

c. Use the formula for the expected value of a discrete distribution to calculate the mean value for the sum of the dice.

d. In the same way calculate the variance and standard deviation of the distribution.

e. What is the relationship between the mean here and the mean in Part 1?

f. What is the relationship between the variance here and the variance in Part 1?

g. What is the relationship between the standard deviation here and the standard deviation in Part 1?

Part 3

Choose one of these hypotheses to be true in your opinion.

a. It is possible that the true probabilities for the rolling of each face are equal.

b. There is significant evidence that the probabilities for the rolling of each face are not all equal.

Your friend says that rolling the die is a random event and we have no way to know what any roll will be. So how can one decide what is the true hypothesis based on this experiment? Construct an argument to convince them that you have chosen the correct hypothesis using probabilities and statistics.

Teacher Notes for A Rolling Die Gathers No Data

Prerequisite Knowledge

- Discrete Probabilities
- Mean and standard deviation of discrete distributions

Equipment needed

- Foam Die
- Box cutter or exacto knife

Objectives

The three objectives of this lab are to calculate the mean and variance of a discrete distribution from a probability table, to verify the summing rule for means and variances for combined events, and as a precursor for the thinking behind hypothesis testing. Groups of 3 or 4 work well for this lab. If done using class time, this activity could be finished in an hour if you do not need a full lab report. If doing this lab outside of class, it is still a good idea to have them shape the dice either in class or show them to you before they collect the data. If you use wooden dice instead of foam then you will need sandpaper. If you use plastic dice you will need a power sander and a pair of pliers. You can prepare the dice before the activity in order to save some time. Students will need 3 to 5 days outside of class to do all the calculations and write the report about what it all means.

Comments and Answers

Part 1

Sample data from a modified die

X	1	2	3	4	5	6
$P(X)$	.32	.15	.17	.14	.16	.06

$$\mu = 1(0.32) + 2(0.15) + 3(0.17) + 4(0.14) + 5(0.16) + 6(.06) = 2.85$$

$$\sigma^2 = 1^2(0.32) + 2^2(0.15) + 3^2(0.17) + 4^2(0.14) + 5^2(0.16) + 6^2(.06) - 2.85^2 = 2.7275$$

The mean is 2.85, the variance is 2.7275, and the standard deviation is 1.6515

Part 2

Data tables for the sample data are given at the end of these notes. The mean is 5.7, the variance is 5.455, and the standard deviation is 2.3356

The connections you hope to find are ...

$5.7 = 2.85 + 2.85 = 2.85(2)$

$5.455 = 2.27275 + 2.27275 = 2.27275(2)$

$2.3356 = \sqrt{1.6515^2 + 1.6515^2} = 1.6515\sqrt{2}$

If X is the roll of one die and Y is the sum of two dice then you can express Y = X + X, so the expected value E(Y) = 2E(X), and the variance V(Y) = 2V(X).

Part 3

We can hope that students will choose the hypothesis that this is not a fair die. An expected value of 2.85 is only slightly less than an expected value of 3.5 for a fair die. After 100 rolls of the die the expected sum of a fair die would be 3.5(100) = 350.

$$Z = X + X + X + ... + X$$

$$E(Z) = 100E(X)$$

$$V(Z) = 100V(X)$$

$$\sigma(Z) = \sqrt{100V(X)} = 10\sigma(X)$$

We have found a sum far less (at 285 for the sample data). And the variance of this sum would be 2.727(100) = 272.7 giving a standard deviation of 16.5 for the sample data. Now our expected value for 100 rolls is 350 - 285 = 65 away and is about than 65/16.5 = 4 standard deviations from the what a fair die should give.

$$\frac{x - \mu}{\sigma} = \frac{350 - 285}{16.5} = 3.939$$

Any value 4 S.D. or more from the predicted value has a very low probability regardless of the distribution. So it is very unlikely that this die is a fair die and that the results were due to random chance.

Table 13.1: Probabilities for rolling two dice

Y	X	1	2	3	4	5	6
X	$P(X)$	.32	.15	.17	.14	.16	.06
1	.32	.1024	.048	.0544	.0448	.0512	.0192
2	.15	.048	.0225	.0255	.021	.024	.009
3	.17	.0544	.0255	.0289	.0238	.0272	.0102
4	.14	.0448	.021	.0238	.0196	.0224	.0084
5	.16	.0512	.024	.0272	.0224	.0256	.0096
6	.06	.0192	.009	.0102	.0084	.0096	.0036

Table 13.2: Probabilities for the sum of the faces on two dice

Y	2	3	4	5	6	7	8	9	10	11	12
$P(Y)$	.1024	.096	.1313	.1406	.1733	.134	.092	.6502	.0424	.0192	.0036

Lab 14
It's in the Cards

Equipment needed

- One deck of cards

Part 1

a. Let each card represent a value in a uniform distribution. Ace = 1, Jack = 11, Queen = 12, and King = 13. What is the mean, variance, and standard deviation of this population of 52 cards? Explain how you found these values.

b. Shuffle the cards and take the top two cards. Find the average of those two values and record it.

c. Repeat taking two cards and finding the average of each pair until you finish the deck.

d. Now find the sample mean, sample variance, and sample standard deviation of the 26 averages.

e. Comment on how those values compare to the values you found in step a.

Part 2

a. Shuffle the cards again and take the top four cards. Find the average of those four values and record it.

b. Repeat taking four cards at a time until you finish the deck. Then shuffle and do it again. You should now have 26 averages.

c. Find the sample mean, sample variance, and sample standard deviation of the 26 averages.

d. Comment on how those values compare to both sets of values you found in Part 1.

e. Make a prediction of what would happen if you were to do this again using sets of eight cards instead of two or four. Explain how you came up with your prediction.

Part 3

a. In this part you will create three plots. To make the comparison easier, limit your population to half of the deck, such as only the red cards, so that all three graphs use 26 values. Now make either a dot plot or a histogram with bin width of one for the population of 26 cards, the sample means of $n = 2$, and the sample means of $n = 4$.

b. Since you have already discussed center and spread in Parts 1 and 2, focus your observations on shape and extreme values. Comment on similarities and differences in these three plots.

c. Make a predictive plot of what would happen if your were to do this again using sets of eight cards instead of two or four. Explain how you came up with your prediction.

Part 4

When you take a single sample, the mean value you find is one element in a larger collection of mean values. Those mean values have their own distribution. It is not the distribution of the population that was sampled, but it is related to that distribution. The three important things we know about the distribution that this mean of a sample belongs to are the following:

1. The meta-mean (the mean of all these means) is equal to the mean of the distribution. $\mu_{\bar{x}} = \mu$

2. The standard deviation of the set of all these means is smaller than the standard deviation of the population by a factor of $\sqrt{n}$. Or you could say the variance of the means is one over n times the variance of the population. $\sigma_{\bar{x}} = \dfrac{\sigma}{\sqrt{n}}, \sigma_{\bar{x}}^2 = \dfrac{\sigma^2}{n}$

3. The shape of the distribution of means is "more normal" than the shape of the population AND increases in "normality" as the sample size increases.

Write a few lines about how your results from this lab either support or contradict these three statements.

Teacher Notes for It's in the Cards

Prerequisite Knowledge

- Finding both population and sample statistics.

Equipment needed

- One deck of cards

Objectives

The objective of this lab is to study the distribution of sample means. We have usually done this activity before teaching the sampling distribution, and introducing that topic a couple of days before the lab is due. Groups of 4 make data collection go pretty quickly. If given class time to collect the data, then you can ensure that students are doing the right thing. Then give the groups 3 or 4 days to finish the lab.

Comments and Answers

Part 1

a. $\mu = \frac{1}{n}\sum x_i = \frac{1}{52}(364) = 7$

$$\sigma^2 = \frac{1}{n}\left(\sum x_i^2 - n\mu^2\right)$$

$$= \frac{1}{52}\left(3276 - 52(7)^2\right) = 14$$

$$\sigma = \sqrt{14} = 3.7417$$

b. For example $(6 + 4)/2 = 5$

c. This may seem like a long slow job, but with 4 students, each taking two cards at a time, the deck is done very quickly. (Especially if students can average the values in their heads.)

d. $\bar{\bar{x}} = \frac{1}{n}\sum \bar{x}_i = \frac{1}{26}(182) = 7.$

$$s^2 = \frac{1}{n-1}\left(\sum \bar{x}_i^2 - n\bar{\bar{x}}^2\right)$$

$$= \frac{1}{25}\left(1440.5 - 26(7)^2\right) = 6.66$$

$$s_2 = \sqrt{6.66} = 2.5807$$

e. Students will note that $\bar{\bar{x}} = \mu$ and that $s < \sigma$. Some may also note that s is not half of σ, but the sample variance is close to half of the population variance.

Part 2

a. $(4 + 13 + 7 + 1)/\ 4 = 25/4 = 6.25$

b. Again, working as a group this goes pretty quickly.

c. $\bar{\bar{x}} = \frac{1}{n}\sum \bar{x}_i = \frac{1}{26}(182) = 7.$

$$s^2 = \frac{1}{n-1}\left(\sum \bar{x}_i^2 - n\bar{\bar{x}}^2\right)$$

$$= \frac{1}{25}\left(1354.62 - 26(7)^2\right) = 3.225$$

$$s_4 = \sqrt{3.06} = 1.796$$

d. Students will note that $\bar{\bar{x}} = \mu$ again and that $s_4 < s_2 < \sigma$. Some may also note that s_4 is about half of σ, or the sample variance is close to half the other sample variance.

e. Students should predict that $\bar{\bar{x}}_8 = \mu$ again and that $s_8 < s_4 < s_2 < \sigma$. The variance and standard deviation should be close to 1.75 and 1.3 respectively but do not expect students to get these values correct.

Part 3

a. The second and third plots are based on sample data at the end of these notes and will be different for each group.

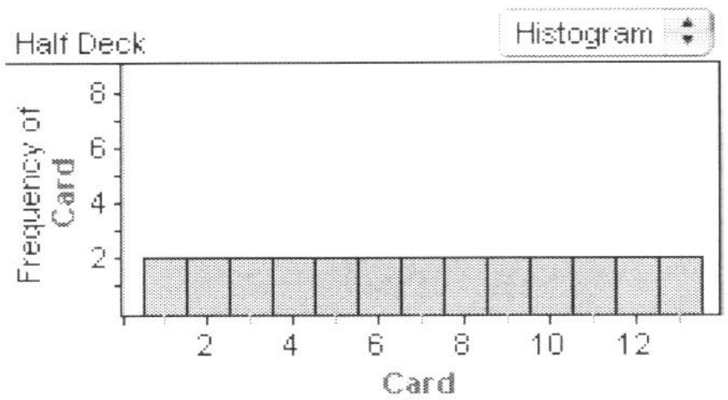

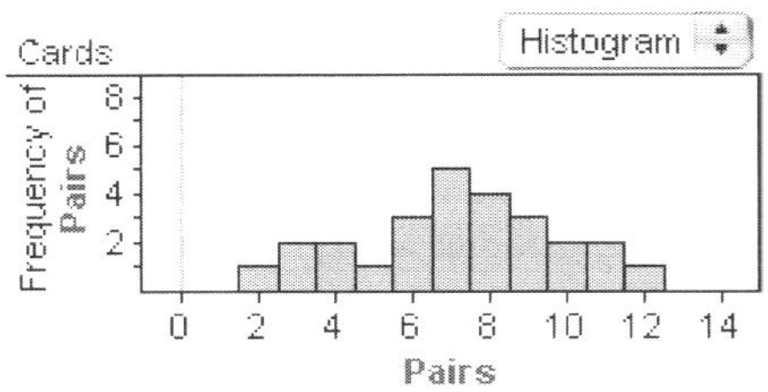

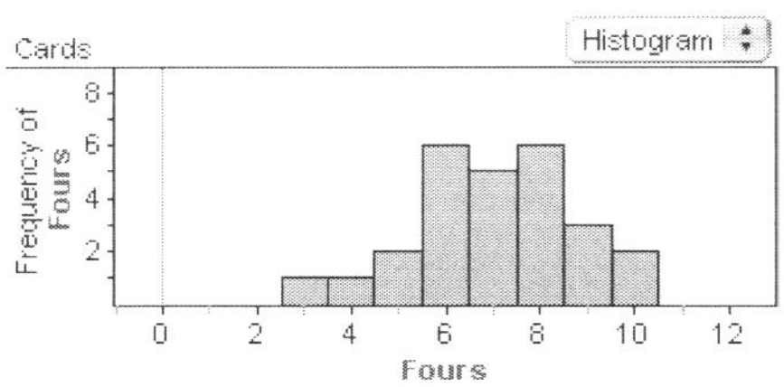

b. As sample size increases, so does the height of the mound. Also, the number of extreme values gets smaller and the number of center values gets larger.

c. This plot should have almost nothing outside of the 4 to 10 range and many values in the 6 to 8 range.

Part 4

Due to the way data collection was done, students should all agree that point one is confirmed. Of course, had we done random sampling we would not have found this to be as exact as we did. The true values for size two distributions should have been $\sigma^2_{\bar{x}} = 7, \sigma_{\bar{x}} = 2.646$ and for size 4 distribution $\sigma^2_{\bar{x}} = 3.5, \sigma_{\bar{x}} = 1.871$. Hopefully students will accept that the values they found are not that far from these values. Certainly they will find their shapes to be more mounded than the original flat distribution. It is not likely that they will have nice symmetric bell curves, but at least in comparison they should see that the distribution looks a little more normal.

Table 14.1: Sample averages

Pairs	8	12	7.5	3.5	9	6	6.5	11	3	2	8.5	7	9.5
	6	3	6.5	10.5	6	7	9	6.5	7.5	5	10	4	7.5
Fours	7.25	5.5	9.25	8	6.5	5.5	8.25	9	6	9	7.75	8	5.25
	5.25	7.25	10.25	5.75	4	7.5	6.75	10.25	3	5.5	7.75	7.25	6.25

Lab 15
The Confidence Game

Finding an interval estimate, or confidence interval, for a mean is typically just evaluating a formula, or better yet, a function built into your technology. Understanding what that interval means and interpreting the interval to the reader is a bit more complicated.

Equipment needed

- One or more decks of cards

Part 1

a. Let each card represent a value in a uniform distribution. Ace = 1, Jack = 11, Queen = 12, and King = 13. What is the true mean of this population?

b. Shuffle the cards and deal out eight cards to each member of the group. Use your eight values to find a 90% confidence T interval for the mean. Record the interval for each group member.

c. Reshuffle and deal again. Repeat this procedure until the group has found a total of at least 50 intervals.

d. Prepare a graph showing all 50 (or more) of the intervals. Your graph should be similar to this example showing the first 4 intervals found by another group.

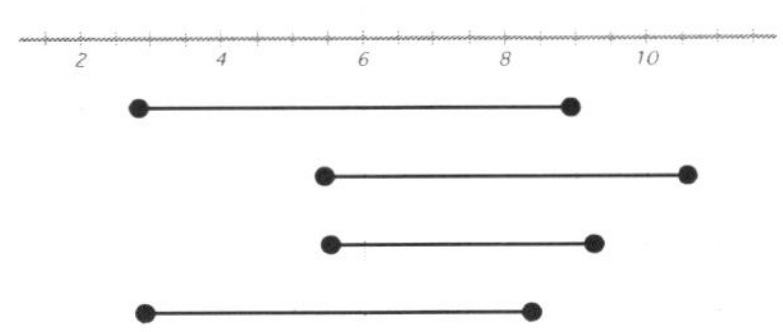

e. Add a vertical line to your graph at the true mean of the population.

Part 2

a. Each interval found in Part 1 is an estimate intended to capture the true mean of the population. What percent of the intervals you found did actually capture that mean?

b. In real life you only take one sample of data and you only construct one interval. Will that interval contain the true mean?

c. You instructor will now provide you with eight cards from a "special" deck. This deck is made from parts of at least three different decks. Record your eight values.

d. Make an 90% interval estimation of the mean value of this deck of cards.

e. Now you must explain to someone the meaning of your interval. You can't talk about this interval with probabilities because it is in the past. But you still need to express your uncertainty that this interval truly contains the mean. What you must make clear is your uncertainty comes from the process or formula. Remember: the true mean is a fact and has no probability, the interval is a fact and has no probability.

Part 3

While it serves us in this lab to provide a detailed, multi-sentence, explanation of the meaning of a confidence interval, in general practice it would be nice if one single sentence was used to explain the meaning of the interval. Each sentence below is an attempt. Some are good, some are poor, and some are just wrong. Make such a classification for each statement and explain what is good, poor, or wrong about each.

a. Ninety percent of all sample means are likely to fall in this interval.

b. Ninety percent of all true means are likely to fall in this interval.

c. There is a 90% chance that the sample mean falls in this interval.

d. There is a 90% chance that the true mean falls in this interval.

e. Ninety percent of all samples like this one will contain the true mean.

f. Ninety percent of all intervals like this one will contain the true mean.

g. We are 90% sure that this interval contains the true mean.

h. We are 90% sure that intervals found this way contain the true mean.

i. Only 10% of all means would not be in this interval.

j. Only 10% of all intervals like this would not contain the true mean.

k. Only 10% of all intervals like this would not contain the sample mean.

l. Only 10% of all sample means would not fall in this interval.

m. Only 10% of all true means would not fall in this interval.

Teacher Notes for The Confidence Game

Prerequisite Knowledge

- Finding confidence intervals for the mean with technology

Equipment needed

- A deck of cards for each group
- Three decks of cards for you.

Objectives

The objective of this lab is to look at the language of interpreting a confidence interval and to understand why our uncertainty lies in the set of all intervals not in the one interval we have created. With groups of 4 students, data collection does not take too long. Even with the collection of 50 samples and finding the confidence intervals for each, this lab can be done in a single class period. Outside of class we would give 4 or 5 days to complete.

Comments and Answers

Part 1

a. This population has a mean of 7.

b. The students should be able to enter the eight values into their technology and request a 90% T interval. They need only to record the interval (i.e. 4.993 to 9.507)

c. Students need to see that some intervals do not contain the mean. This should be about one in ten, so with 50 intervals usually each person has one such interval.

d. This will take some planning to get all 50 on the same graph. Students should think before they start drawing. Note: eleven inch graph paper ruled 5 to the inch will have 55 lines top to bottom.

e. This line should hit 45 of the 50 intervals. You might have students highlight the intervals that do not contain the population mean.

Part 2

a. While it will likely not be exactly 90% that capture the true mean, it should be close. If not, they may be doing the intervals wrong. The worst case we found when doing everything correctly was 84%. We have also, on very rare cases, had 100%.

b. The interval will likely contain the true mean but you have no way to know.

c. The "special" deck used here should be composed of either mostly all high cards or mostly all low cards. Just get three decks all the same and sort cards into two piles. It is good to keep a few low cards in the "high" deck and a few high cards in the "low" deck, but not too many. Depending on how many groups you have, you may need only one of the decks, or you may need both. You can find the true mean of your two decks but it isn't really that important and your students should definitely not know the true mean. They should record the actual 8 cards they were given so that you can check this interval. This may tell you if the first intervals were found properly.

d. It is good for them to show some work here such as the sample mean and sample standard deviation.

e. Probability, as defined in this course, is a likelihood of some future random event meeting a given criteria. Past events and facts have no probability, they are either true or false. Often people want to express a meaning to an interval that they found using probability language. But once the sample is taken, it is in the past. This means that once you find an interval you cannot talk about the probability that THAT interval contains the true mean. Either it does or it does not. Thus it is a fact and has no probability. Of course you typically have no way of knowing the true mean, so you can't say if the interval does or does not contain it. So you remain uncertain of the truth. This is the problem. If students haven't been "taught" how to interpret an interval, most students will use two or three sentences to explain what this interval tells them. When they ask for our help we try to get them to write what they know about the whole process and how the 90% comes into the picture.

Part 3

a. False. The percent of sample means falling into this interval will be 90% only if the sample mean used to create the interval is equal to the true mean. For most intervals the percent of sample means will be much smaller.

b. False. There is only one true mean, thus it must be 0% or 100%.

c. False. The interval is constructed around the sample mean thus it is 100% sure to be in the interval .

d. False. See b

e. False. The sample is a set of 8 values. The true mean may not even be an integer thus having no chance to be in the sample.

f. Poor. While true, it does not tell you specifically about this interval.

g. Good. This is the one liner we most often see. It gets its meaning across without making a probability out of it.

h. Poor. See f.

i. False. See a.

j. Poor. See f.

k. False. See c.

l. False. See a.

m. False. See b.

Lab 16
p is for Pepper

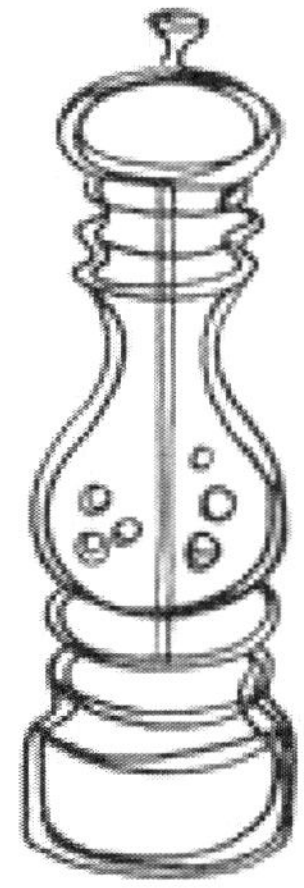

Two chefs were locked in a kitchen. When they were released, each one insisted that he had been held captive and forced to eat food made by the other chef. Who is telling the truth?
Thirty-six different food dishes were found in the kitchen. The foods were analyzed and found to have a mean black pepper concentration of 47 ppm (with a standard deviation of 12 ppm). All of the dishes were actually prepared by the same chef. The personal kitchens of each chef were also tested. Chef Green's dishes had a mean pepper content of 49 ppm, while Chef Redd's dishes had a mean of 42 ppm. (You will use the sample standard deviation for the pepper levels of 12 ppm for all parts of this lab.)

Part 1

It looks like Chef Green might have done the cooking, but let's calculate some probabilities.

a. What is the probability that the 36 items would have a mean of exactly 47 if the true mean were 42?

b. What is the probability that the 36 items would have a mean of exactly 47 if the true mean were 49?

c. What is the probability that the 36 items would have a mean of more than 47 if the true mean were 42?

d. What is the probability that the 36 items would have a mean of more than 47 if the true mean were 49?

e. What is the probability that the 36 items would have a mean of less than 47 if the true mean were 42?

f. What is the probability that the 36 items would have a mean of less than 47 if the true mean were 49?

g. Of these six probabilities, which is the one that points most clearly to which chef did or did not do the cooking? Explain why.

Part 2

Match each probability found in Part 1 with one of these statements. Explain your choice including a sketch of a shaded normal distribution representing the probability.

a. The probability of this level of pepper if Chef Green did the cooking.

b. The probability a pepper level is in an interval close to his mean if Chef Green did the cooking.

c. The probability a pepper level is not in an interval close to his mean if Chef Green did the cooking.

d. The probability of this level of pepper if Chef Redd did the cooking.

e. The probability a pepper level is in an interval close to his mean if Chef Redd did the cooking.

f. The probability a pepper level is not in an interval close to his mean if Chef Redd did the cooking.

Part 3

Make the correct word choices for each statement.

a. To show that Chef Green did the cooking we found a [large]/[small] probability when we assumed [Chef Redd]/[Chef Green] was cooking.

b. A [high]/[low] probability convinces us that an assumption must be [wrong]/[right].

c. In order to show a statement is true we should first assume the [statement]/[opposite] is true.

d. Then look for evidence that this is [unlikely]/[likely].

e. Once you assume something is true then it is [hard]/[easy] to use probability to prove it, but [hard]/[easy] to use probability to disprove it.

In your own words explain what this lab has to do with hypothesis testing and p-values.

Teacher Notes for p is for Pepper

Prerequisite Knowledge

- Probability of sample mean intervals
- At least one hypothesis test

Objectives

The objective of this lab is to understand what the $p-$value tells you and why we start with the null hypothesis. This is a shorter lab. It can be done by any size group or even an individual. You can do the lab in less than one class period or a day out of class. There is no hypothesis test in the lab, and students could complete the entire lab except for the very last question without knowing how to do a hypothesis test. We use this lab after we have spent a week on testing, but you could also use it to introduce the concept of p-value.

Comments and Answers

Part 1

a. Two different answers for this question are acceptable. The probability of a single value in a continuous distribution is given to be zero, so that is the first and most common correct answer. But we can also see this as the probability that to the nearest ppm the mean of 47 is actually the probability that the mean lies between 46.5 ppm and 47.5 ppm. Converting these to T values gives $\frac{46.5-42}{12/\sqrt{36}} = 2.25$ and $\frac{47.5-42}{2} = 2.75$. The probability is then $P(2.25 < T < 2.75) = 0.011$ when there are 35 degrees of freedom.

b. Again we have accepted either 0 or $P(-1.25 < T < -0.75) = 0.119$

c. $P(T > \frac{47-42}{2}) = P(T > 2.5) = 0.009$

d. $P(T > \frac{47-49}{2}) = P(T > -1) = 0.838$

e. $P(T < \frac{47-42}{2}) = P(T < 2.5) = 0.991$

f. $P(T < \frac{47-49}{2}) = P(T < -1) = 0.162$

g. Neither a or b tell us anything. These probabilities are too small to convince us that something is true and too close to tell us one is better than another. Answer c is also a low probability which indicates that this is unlikely. Both d and e are large probabilities telling us both are likely so they are no help in making a decision. Finally f is small, but not that small that we would find it to be unlikely. So the most convincing is that a value like 47 or more is very unlikely for a true mean of 42 so it is probably not Redd.

Part 2

a. Chef Green has a mean of 49 and since this is a single value it matches b.

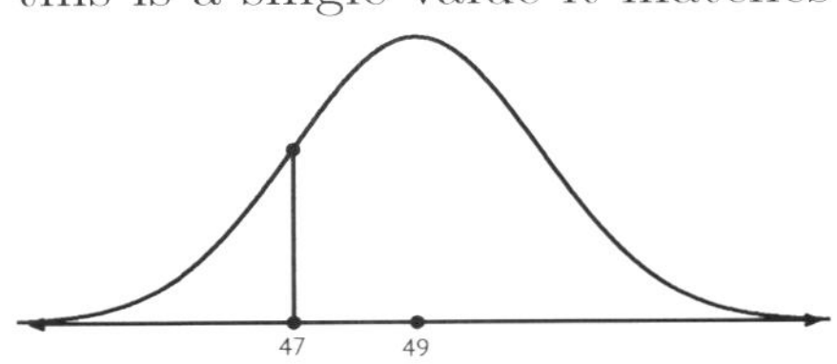

b. The interval close to 49 would be greater than 47 so this matches d.

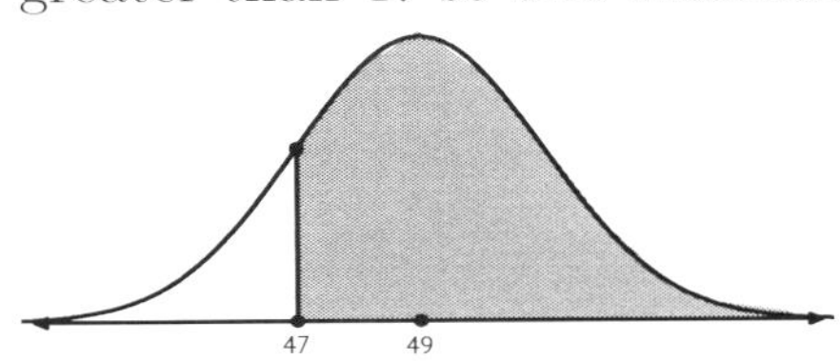

c. The interval not close to 49 would be less than 47 so this matches f.

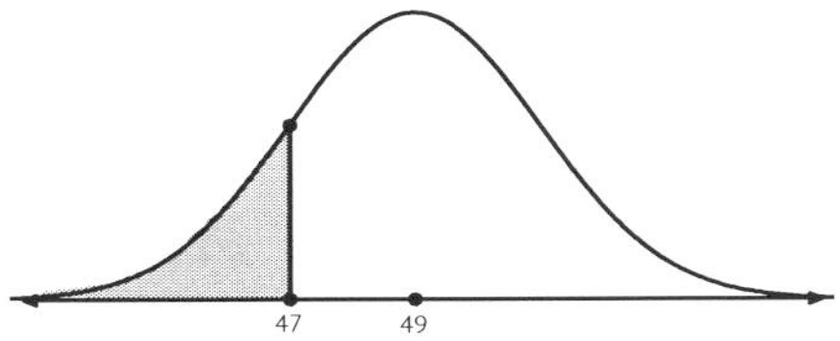

d. Chef Redd has a mean of 42 and this is a single value so it matches a.

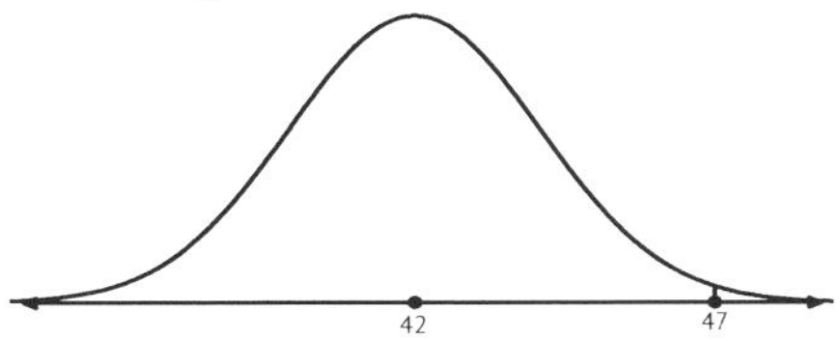

e. The interval close to 42 would be less than 47 so this matches e.

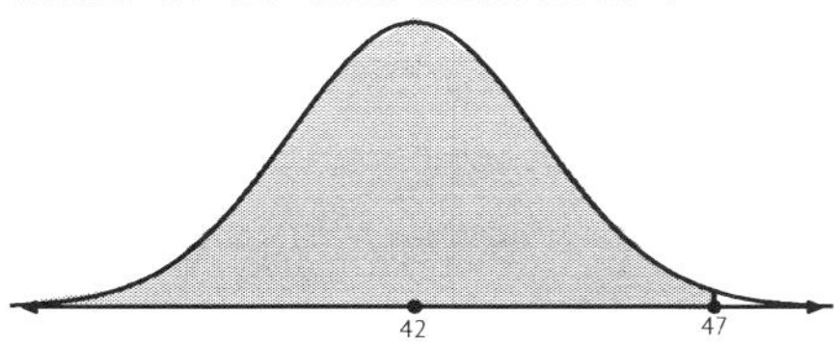

f. The interval not close to 42 would be greater than 47 so this matches c.

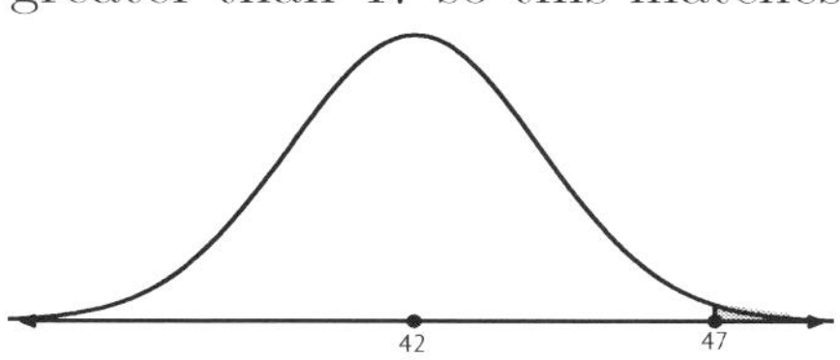

Part 3

a. To show that Chef Green did the cooking we found a **small** probability when we assumed **Chef Redd** was cooking.

b. A **low** probability convinces us that an assumption must be **wrong**.

c. In order to show a statement is true we should first assume the **opposite** is true.

d. Then look for evidence that this is **unlikely**.

e. Once you assume something is true then it is **hard** to use probability to prove it, but **easy** to use probability to disprove it.

In hypothesis testing we calculate a conditional probability, given the opposite of what we believe. Then we hope to show this is very unlikely. This leads us to reject our hypothesis. Once that is rejected, we are left showing that our belief was correct.

Lab 17
Insignificantly Significant

One problem that can happen in any test of significance is called oversampling. In this lab you will look at the most basic tests; a one sample T test for the mean of a distribution, and a one proportion test, to see how this problem affects your interpretation of results.

Part 1

a. Use your technology to generate a sample of 50 random values from a normal distribution whose true mean is 80 and standard deviation is 20.

b. Now perform a T test with the alternate hypothesis that the true mean is not equal to 80. With type one errors and all you will most likely find, with 95% confidence, that you fail to reject the null hypothesis. Give your sample mean and standard deviation $(\bar{x}, s_x)$ as well as the T and p-value from the test.

c. Now increase your sample by adding a second set of fifty values to your list that are identical to the first fifty. What are the new mean and standard deviation of this sample of 100? Explain why.

d. Perform a T test with the sample of 100 and give the T and p-values for this test. Explain why these values have changed.

e. If you were to continue adding identical sets of fifty values again and again, how big would the set have to be before the test produced a p-value below 0.05? Explain your work leading to this sample size.

f. This is called oversampling. You know the alternate hypothesis is false and the null hypothesis is true but yet we are told to reject the null by the small p-value. Explain why this happened.

Part 2

An group of eager young statisticians wants to make a name for itself (and show their NFL team was robbed of victory). They get the coin used to determine the start of the Superbowl game. After several long days of flipping, they announce to the world that this coin was not fair and they have used statistics to prove it is biased towards "Heads". The results claimed by the statisticians are that after 100,000 flips there were 50,284 heads and 49,716 tails. Give the results of the test "proving" the proportion of heads is greater than 50%, and then explain why you shouldn't be to concerned. Why are the terms "statistical significance" and "practical significance" are not the same thing?

Part 3

Find a article in a research magazine or e-zine that sites statistical results from a very large sample and comment on the conclusions reached in the article. You often find this type of study done in the social or soft sciences.

Teacher Notes for Insignificantly Significant

Prerequisite Knowledge

- One sample T-test
- One sample proportion test

Objectives

The objective of this lab is to understand the dangers of oversampling, and conclusions reached for very large samples. Groups of 2 up to 4 can work in this lab. Part 3 may take some time. If you include that, you may need to give them a week to complete the report.

Comments and Answers

Part 1

a. On TI-84: randNorm(80, 20, 50) $\rightarrow L_1$

b. Answers will vary. For example, $\bar{x} = 78.02, s = 21.73, T = -0.6443$, p-val $= 0.5224$

c. On TI-84: augment$(L_1, L_1) \rightarrow L_2$. The mean will be the same and the standard deviation will be essentially the same.

d. $T = -0.9150, p\text{-val} = 0.3620$, As n is now larger the standard error is smaller so $\bar{x}$ falls further from the null mean as measured by standard errors.

e. Repeating augment$(L_1, L_2) \rightarrow L_2$ and testing. When $n = 500$ then $T = 2.056, p\text{-val} = 0.0403$.

f. When the sample is very large the standard error becomes very small. So even a small difference in the mean will be statistically significant and result in a low p-value.

Part 2

Again test results give $Z = 1.796$ and p-val $= 0.0362$ on a one sided test. But the standard error is $\sqrt{(0.5)(0.5)/100000} = 0.0016$. So even though the proportion is less than 0.3% away from 50%, this is enough to show statistical significance. But the coin is still fair and it is just as likely that the next 100,000 tosses might show the coin to be tails biased. So we say this is not practical significance because we are looking at minor differences to prove the point.

Part 3

Depending on what your school library has access to, it may be difficult to find these articles if such journals are not part of your circulation. This oversampling actually happens quite frequently. It doesn't mean the test is bad or wrong, it is just harder to determine if the significance found is practical or just statistical. What students need to look for are articles where the writers have used very large sample sizes such as using hundreds or even thousands of measures in a hypothesis test.

Lab 18
A Safe Prediction

Equipment needed

- Rubber bands
- Water bottle
- Meter stick or tape measure
- (opt) CBR or video camera

Part 1

a. Choose an integer at random between 2 and 12. Build a rubber band chain of this length. This is a mini-bungee jumping cord.

b. Now you need a weight such as a half-filled water bottle to serve as your subject. Attach the weight to your chain. Fix the other end of the bungee to a rigid point and hold the weight at the same height as this point of attachment.

c. Make exactly 4 drops with this set up. For each drop record the maximum distance that the cord stretches. This can be done "by eye", but if equipment for more accurate measurements is available, then use this to make the values as accurate as you can. This might mean placing a CBR on the floor below your drop or setting up a video capture of the drop so you can trace the lowest point for each of the four drops.

d. Repeat steps a, b, and c nine more times (it is OK to get the same random number twice). You should now have a set of 40 ordered pairs ($ChainLength, MaxDrop$).

e. Make a scatter plot of your data.

f. Find a model and a measure of the association between the two variables, chain length and maximum drop.

g. Note that you will need this exact water bottle with the same amount of water in Part 3 so do not lose it or change it in any way.

Part 2

Based on the data you have collected you will design an exciting, but safe, drop in a new location with a known maximum height. This height will be provided by your instructor. This value will be called d in this lab.

a. Even if you knew the exact model for your data ($\mu_y = \alpha + \beta x$) this would only predict the average maximum stretch and there would still be the natural variation in the experiment, σ_e^2. Estimate this value using MSE. This is your first source of variation, find this now.

$$s_e^2 = \frac{1}{n-2}\sum_{i=1}^{n}(y_i - \hat{y}_i)^2$$

b. You don't really know the exact model, so the mean height is is also an interval. This is your second source of variation. Fortunately you know about the distribution of the sample mean, and you know that that variation is just one over n times the actual variation. Find this variance now.

$$s_{\bar{y}}^2 = \frac{s_e^2}{n}$$

c. These two values make up the variation of a prediction that is correct in the center of your domain. But as you move out from the center of your x-values (chain length), you have more uncertainty. In this case you are actually making a prediction well beyond the domain you used to collect data so this last bit of variation is significant. It depends on two things: how far you are from the center of your domain and how wide that domain is. Let x^* represent the chain length you wish to make a prediction for. Then the distance from center is $(x^* - \bar{x})$. Then this is turned into a scale factor with the variance of the x-values. As a scale factor it must be unitless. While there are many ways to do this, the correct way is to divide $(x^* - \bar{x})^2/(n-1)$ by the variance of x. Thus this last source of variation is

$$s_{x^*}^2 = s_e^2 \frac{(x^* - \bar{x})^2/(n-1)}{s_x^2}$$

You don't know the value of x^* (that is our goal) so in this step calculate this variance for any three chain lengths you wish.

d. These values are all independent so the sum is the total variance of your prediction

$$s_{pred}^2 = s_e^2 + s_{\bar{y}}^2 + s_{x^*}^2$$

You find a prediction interval in the same way you find most confidence intervals.

$$\widehat{pred} = a + bx^* \pm \left(t_{\alpha/2,df} \cdot s_{pred}\right)$$

If you define "safe" and 95% confident then you must now find the nearest integer value for x^* that gives an interval that does not contain d. Explain (show) how you found this value.

e. If you define "very safe" as 99% confidence, you may end up with a different x^*. Find this value also and give the prediction interval.

f. Finally if you define "very very safe" as 99.9% confidence, what value and interval will satisfy this condition?

Part 3

This is a competition, your group's score will be the measure of excitement in the drop, measured by the closest distance from the floor when dropped from the announced location. You are only allowed one drop. If the bottle touches the floor then the measure is infinity. The lowest score wins the competition.

Teacher Notes for A Safe Prediction

Prerequisite Knowledge

- Linear regression
- Mean square error
- Confidence intervals

Equipment needed

- Rubber bands
- Water bottle
- Meter stick or tape measure
- (opt) CBR or video camera

Objectives

The objective of this lab is to work with linear models and prediction intervals. This lab is typically done with groups of 4. Depending on how much class time you are willing to give to this activity, this can take anywhere from one to two weeks to complete

Comments and Answers

Part 1

The rubber bands can be attached to the neck of the water bottle with a short piece of wire or a twist tie. It worked well to video record the drops in groups of 4 with a tape measure taped to the wall. Then the videos were transferred to a computer and advanced by frame to the low point of the drop. The CBR is sensitive to movements in and out of the pickup. You will need to experiment and refine the process to get good data. Make sure the values from the CBR agree with an estimate of the drop done with a tape measure.

The answers that follow are based on the sample data found at the end of these notes.

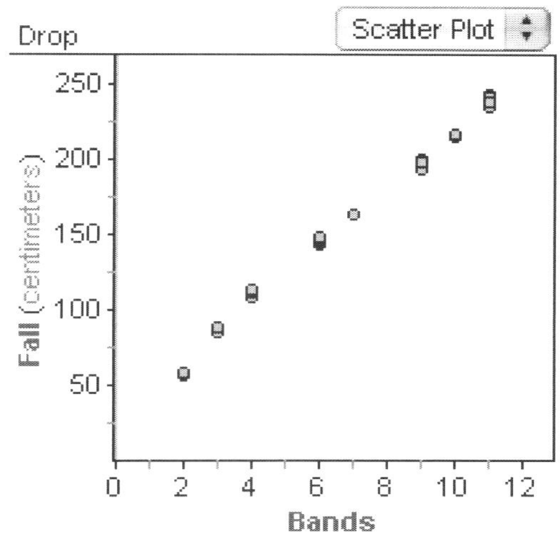

$$\hat{y} = 27.319 + 19.1084x, r = 0.9975$$

Part 2

a. The mean square error is $s_e^2 = 19.5504cm^2$

b. The variance in the mean prediction is $s_{\bar{y}}^2 = 19.5504/40 = 0.48876cm^2$

c. The additional variance at 10, 20 and 30 bands is

$$s_{10}^2 = 19.5504\frac{(10-7.2)^2/(39)}{10.2154} = 0.3847cm^2$$

$$s_{20}^2 = 19.5504\frac{(20-7.2)^2/(39)}{10.2154} = 8.0400cm^2$$

$$s_{30}^2 = 19.5504\frac{(30-7.2)^2/(39)}{10.2154} = 25.5097cm^2$$

d. Finding $t_{0.025,38} = 2.0244$ and the expression

$$s_{pred}^2 = 19.5504\left(1 + \frac{1}{40} + \frac{(x^*-7.2)^2}{39 \cdot 10.2154}\right)$$

then test values in

$$\widehat{pred} = 27.319 + 19.1084x^* + 2.0244\sqrt{s_{pred}^2}$$

to find the largest value of x^* that is less than d. This is the upper limit of the prediction interval. If $d = 5$ meters then $x^* = 24$ gives an upper limit of 497.71 cm.

e. Change $t_{0.005,38} = 2.7116$ and the expression is equal to 482.22 cm when $x^* = 23$ bands.

f. Change $t_{0.0005,38} = 3.5657$ and the expression is equal to 487.07 cm when $x^* = 23$ bands.

Part 3

You want this to be a big drop. At least twice what they have worked with previously. A stair well, off the back of the bleachers, or a balcony are all good choices. We usually set up a camera to capture the drop so that we can judge who wins the contest. In the example done here, the height was 5 meters. Here you will find which groups are gamblers and which are not. The winners and the losers both are the ones who gamble, the very safe and very very safe end up in the middle.

Table 18.1: Sample Data: Fall is measured in cm

Bands	Fall	Bands	Fall
6	143	9	196
6	145	9	197
6	146	9	193
6	147	9	198
10	214	3	86
10	214	3	88
10	215	3	85
10	216	3	87
2	57	4	108
2	56	4	111
2	58	4	112
2	58	4	112
9	197	11	237
9	197	11	241
9	199	11	242
9	200	11	241
11	235	7	163
11	238	7	162
11	238	7	163
11	238	7	163

Lab 19
Is That Die Still Rolling?

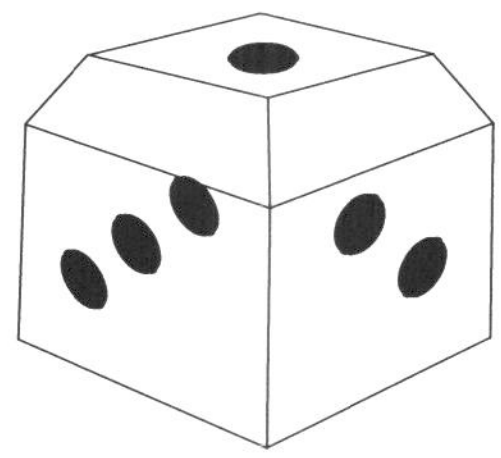

Equipment needed

- Foam Die
- Box cutter or exacto knife

Part 1

If you did the Rolling Die Lab and still have that data then you may skip Part 1.

a. Trim the four edges around the one pip face of the die as evenly as possible. Do not trim any other edges.

b. Roll this die 100 times recording the number of pips on the top face for each roll.

c. Summarize your results in a table.

Part 2

In this lab we will treat these results as categorical data. Imagine replacing the pips on each side with different shapes or different colors. The only relationship between a ⚁ and a ⚃ is that they are different. In this Part we will make our data set binary. The two values are ⚀ and NOT ⚀ (that is, rolling either ⚁ ⚂ ⚃ ⚄ or ⚅)

a. What proportion of your rolls are ⚀?

b. Perform a single proportion test to test the hypothesis that the true proportion is not $\frac{1}{6}$. Record the Z value and the p-value of that test. Show the formula for the Z and the values used.

c. Another way to analyze this data is to compare the observed counts to hypothetical expected counts. Complete this table using what you would expect if it were a fair die. (Remember that expected counts don't have to be whole numbers, but the total for expected counts has to be the same as the total for observed counts.)

	Observed count	Expected count
⚀		
Not ⚀		

d. The difference between what you observe and what you expect is a kind of a deviation. So any measure of variation will look at this difference. Calculate the following statistic, showing all of your work.

$$\frac{(O_1 - E_1)^2}{E_1} + \frac{(O_2 - E_2)^2}{E_2}$$

e. How does the value you found in step d compare to the square of the Z you found in step b?

Part 3

a. While Z works fine when there are two things to compare, what happens when there are more than two proportions? Complete this new table using what you would expect if using a fair die. (Remember that expected counts don't have to be whole numbers, but the total for expected counts has to be the same as the total for observed counts.)

	Observed count	Expected count
⚀		
⚁		
⚂		
⚃		
⚄		
⚅		

b. Calculate this statistic as you did with the simpler version in Part 2, showing all of your work.

$$\sum_{i=1}^{6} \frac{(O_i - E_i)^2}{E_i}$$

c. This is similar to the square of a Z value and probability can be calculated on a distribution. However, having 6 terms in the sum is different than having only two. When there are only two terms we say there is one degree of freedom. (If you know one count you can find the other count because the total is fixed.) When you have 6 terms then there are five degrees of freedom. (You can figure the last count when you know the first five, which means the last term is not independent.) The expected value of this statistic is 1 when there is one degree of freedom and it is 5 when there are five degrees of freedom. Thus we are suspicious when the statistic has any value greater than the degrees of freedom. How does your value compare to 5?

d. The actual distribution followed by this statistic is called the chi-squared (χ^2) distribution. The ch in chi is pronounced "K" as in Christmas. The formula for this distribution is complex (pronounced "ugly") but fortunately, this function is built into most technology. You simply define the lower and upper limits as well as the degrees of freedom and the technology will calculate the area. This statistic is never negative so the distribution starts at zero. Determine how to use your technology to find the area from the value you found in step b up to infinity. Alternatively you can find one minus the area from zero up to your statistic. This is a graph of the χ^2 distribution with 5 degrees of freedom.

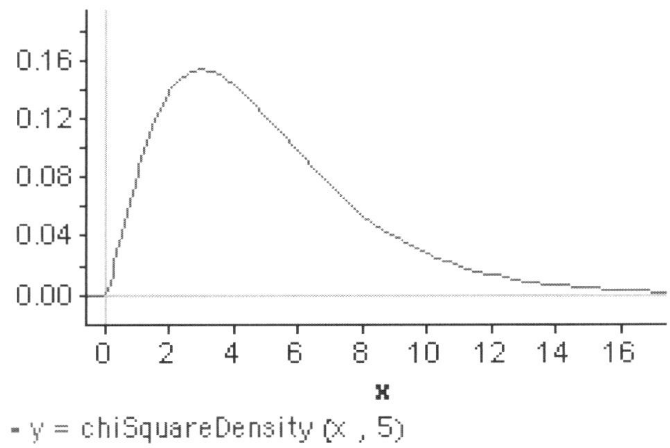

e. What you have done here is called a "goodness of fit" test, where we compare the counts or proportions found in a sample to a hypothesis of a set of other proportions. When this test produces a significantly low p-value, then we must reject at least one of the set of proportions. It is possible that more or even all of the proportions are wrong, but this test only tells us that one proportion is wrong. What statement can you make about your results that reflects this conclusion?

f. The six terms we added together to find our statistic are called the contributions. When any of these contributions is much larger than one, it is evidence that that particular proportion may be the one that is incorrect. There is a danger to using this test when one of the expected counts is too small (like 5 or less). Explain why this is a concern. Create a set of numbers as an example and use it in your explanation.

Teacher Notes for Is That Die Still Rolling?

Prerequisite Knowledge

- Single Proportion Z test

Equipment needed

- Foam Die
- Box cutter or exacto knife

Objectives

The objective of this lab is to understand the goodness of fit test and how it relates to the proportion test. If the group has done the earlier rolling die lab they can reuse that data and this becomes a short lab. If they are making dice and collecting data then groups of 3 or 4 work well for this lab. If done using class time this activity could be finished in an hour if you do not need a full lab report. If doing the lab outside of class it is still a good idea to have them shape the dice either in class or show them to you before they collect the data. If you use wood dice instead of foam, then you will need sandpaper. If you use plastic dice, you will need a power sander and a pair of pliers. You can prepare the dice before the activity to save some time. It still will take 3 to 5 days outside of class to do all of the calculations and write the report about what it all means.

Comments and Answers

Part 1

Sample data from a modified die

Face	⚀	⚁	⚂	⚃	⚄	⚅
Count	32	15	17	14	16	6

Part 2

a. The proportion of ⚀ in the sample data was 0.32

b. $Z = (0.32 - 0.1667)/0.03726 = 4.114$ and p-value $= 0.000\,039$

c. This table uses the sample data

	Observed	Expected
⚀	32	16.67
Not ⚀	68	83.33

d. Using sample data

$$\frac{(32-16.67)^2}{16.67} + \frac{(68-83.33)^2}{83.33} = 16.928$$

e. The statistic should be equal to the square of the Z.

Part 3

a. Using the sample data

	Observed	Expected
⚀	32	16.67
⚁	15	16.67
⚂	17	16.67
⚃	14	16.67
⚄	16	16.67
⚅	6	16.67

b. The total for the sample data is 21.56

$$\frac{(32-16.67)^2}{16.67} = 14.1067$$

$$\frac{(15-16.67)^2}{16.67} = 0.1667$$

$$\frac{(17-16.67)^2}{16.67} = 0.0067$$

$$\frac{(14-16.67)^2}{16.67} = 0.4267$$

$$\frac{(16 - 16.67)^2}{16.67} = 0.0267$$

$$\frac{(6 - 16.67)^2}{16.67} = 6.8267$$

c. 21.56 is much larger than 5, so we can clearly suspect this is not a fair die.

d. χ^2cdf(21.56, ∞, 5) = 0.000 635

e. Because of the low p-value we can conclude that at least one of the proportions is not 1/6.

f. This will be difficult for the students to think about and create. They should create a small set of observed counts and expected counts. If all the observed counts differ from the expected by 5 or more, this does not have much effect when we divide the square of this difference by a large expected count. But if we are dividing by a small value such as 3 or 4, the result will be larger, 6 or 8 or more, and we end up with a large sum giving a misleadingly small p-value.

	Observed	Expected
A	27	32
B	37	32
C	27	32
D	9	4

$$\frac{(27 - 32)^2}{32} = 0.78$$

$$\frac{(37 - 32)^2}{32} = 0.78$$

$$\frac{(27 - 32)^2}{32} = 0.78$$

$$\frac{(9 - 4)^2}{4} = 6.25$$

$$P(\chi^2 > 8.59)_{df=3} = 0.035$$

Lab 20
Jumping *F*rogs and *T* for Two

Equipment needed

- Several sheets of paper of differing weights or thickness
- Directions for folding origami jumping frogs

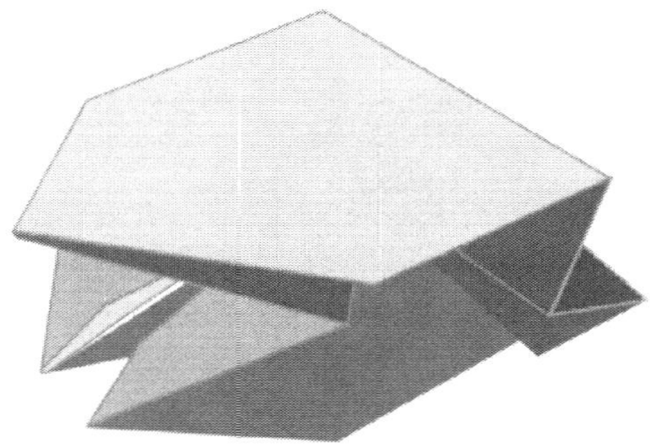

Part 1

How much does the choice of paper determine the distance an origami jumping frog will jump?

a. Design an experiment to test which type(s) of paper are the best and worst for jumping frogs. Remember that any variable or factor, other than paper type, that can't be kept constant, must be randomized. Record the details of your design.

b. Perform your experiment as you have designed it. Collect enough data so that you feel you know if the paper type does make a difference or not.

Part 2

In this part you will select two of the paper types and the data collected from those two. Put aside all the data from any other paper types.

a. What are the mean jumping distances for these two paper types?

b. Perform a two sample T test to compare the means of each paper type. Use a two sided pooled test. Record the T-value and p-value. Show the formula for the T and the values used.

c. Another way to analyze this data is to treat these two samples as a single data set and to look at different sources of variation. Here there are two sources; the variation in jumping distance that has nothing to do with the type of paper and the variation that does. As a single set of data, what is the mean of all the jumping distances? The three means will be referred to as $\bar{x}_1, \bar{x}_2$ and $\bar{x}$ for the mean of all.

d. Create a 9 column table either on paper or on your technology. In Column 1 place the data. In Column 2 place either $\bar{x}_1$ or $\bar{x}_2$ according to which set the data is from. Calculate the values for the remaining columns.

1	2	3	4	5	6	7	8	9
x	$\bar{x}_i$	$\bar{x}$	$x - \bar{x}_i$	$\bar{x}_i - \bar{x}$	$x - \bar{x}$	$(x - \bar{x}_i)^2$	$(\bar{x}_i - \bar{x})^2$	$(x - \bar{x})^2$

e. Sum all 9 columns. What relationships can you find with these sums? (Look at them in three groups of three.)

f. The sums of the last three columns are referred to as sums of squared deviations. If you divide these sums by the degrees of freedom you get the variance. Divide the sum of the last column by $n - 1$ and compare this to the variance of the original column.

g. Degrees of freedom are conserved, and this set of data has $n - 1$ of them. The degrees of freedom for column 8 is one. This is because there are two groups, or two types, of paper and $2 - 1 = 1$. The degrees of freedom for column 7 is $n - 2$, so that the degrees of freedom for columns 7 and 8 sum to that of column 9. Find the variances for deviations in columns 7 and 8 by dividing those column sums by their degrees of freedom. The variance from column 7 is called MSerror (mean square error) and is the variation that has nothing to do with the paper type. The variance from column 8 can be called MSpaper and is the variation due to the type of paper used. For reasons you have already shown, the variance from column 9 is called MStotal.

h. Divide MSpaper by MSerror and label this value F. How does this statistic compare with the square of the T value you found in step b?

Part 3

While a T test works fine when there are two groups to compare, what happens when there are more than two paper types? You could do a separate T test for each pair of paper types, but every test that is done will increase the overall probability of type I and type II errors. So it is best if you use only one test to compare all the means. The ratio of two independent variances (F) is a way to do that. Because of the connections you have observed, there is quicker way to get this ratio than all the steps you did in Part 2.

a. You need only two lists. The first list contains of all the data. The second list has the mean for the group the data value is from (as in columns 1 and 2 in Part 2). Find the variances of each of those lists and label them V_1 and V_2. Also label n the total number of data values, and a the total number of paper types.

b. Use the formulas here and show your work.

$$MSpaper = \frac{(n-1)V_2}{a-1}, \; MSerror = \frac{(n-1)(V_1 - V_2)}{n-a}, \qquad F = \frac{MSpaper}{MSerror}$$

c. If the error from paper was the same as the normal jumping error, that is if $F = 1$, then there is no reason to believe that the type of paper has any effect on the jump distance. But when F is more than one we have evidence that at least one of the paper types makes a difference. What does your value of F tell you?

d. The actual distribution followed by this statistic is called the F distribution. The formula for this distribution is complex (pronounced "ugly"), but fortunately this function is built into most technology. You simply define the lower and upper limit as well as the upper and lower degrees of freedom and the technology will calculate the area. This statistic is never negative so the distribution starts at zero. Determine how to use your technology to find the area from the F value you found in step b up to infinity. Recall the the numerator has $a - 1$ degrees of freedom and the denominator has $n - a$ degrees of freedom. This is a graph of the F distribution with 3 and 16 degrees of freedom.

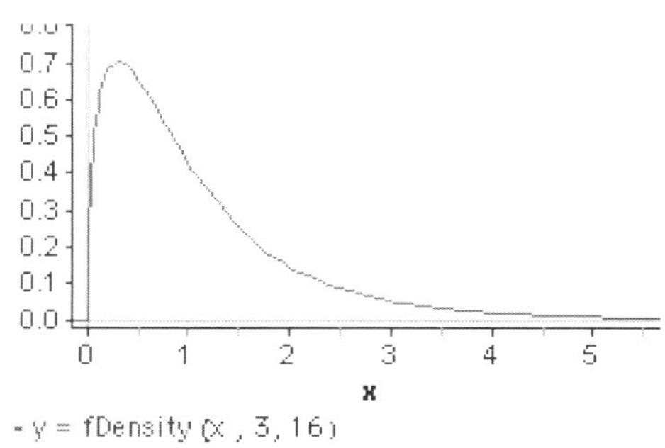

e. What you have done here is called an analysis of variance test, or ANOVA, where we compare the sample means to a hypothesis that states all of the true means are equal. When this test produces a significantly low p-value then we must reject that all of the means are equal. It is possible that more or even all of the means are different, but this test only tells us that one mean is out of sync. What statement can you make about your results that reflects this conclusion?

f. The analysis can be extended to determine which mean(s) is/are different, but usually you can just look at the sample means and figure out that part. What do the sample means in this case tell you about the paper types and jumping distance?

Teacher Notes for Jumping *F*rogs and *T* for Two

Prerequisite Knowledge

- Two sample pooled T test

Equipment needed

- Paper of at least three weights (preferably 4)
- Directions for folding frogs

Objectives

The objective of this lab is to understand that the ANOVA test is an extension of a two sample T test of means to be used when there are more than two groups. This lab is long, 7 to 10 days to complete. You can cut this in half by skipping Step 1 and providing them with sample data. Used at the end of the year it is a good way to bring back elements of experiment design.

Comments and Answers

Part 1

While any size paper will work the smaller the frog the better the jump. A 3 inch by 5 inch works great but we usually cut a full sheet in quarters and use a 4.25 by 5.5 sheet to make our frogs. There are several types of origami jumping frogs, from easy, taking less than 10 folds to create, to more complex versions. Directions can be found in many locations. As long as all the frogs are made are the same, then the origami skill level of the students may determine the type of frog. You may wish to provide the instructions, or let students find their own on the internet. We usually have one or two groups share their experiment design for the rest of the class to critique and offer suggestions.

Sample Design: Make 5 frogs from each of the four weights of paper and number the frogs 1 to 20. Have the same person do all of the folding. Randomly sort the numbers 1 to 20 using a calculator. Jump each frog in the order given by the calculator. Have the same person jump each frog. Measure the distance jumped. Have the same person do all the measurements.

Part 2

These sample answers come from the data at the end of the notes using the lightest and heaviest paper.

a. Mean distance for 20 lb is 10.8 cm and mean for 32 lb is 21.8 cm.

b. The pooled standard deviation is

$$s_p = \sqrt{\frac{4(1.9235^2) + 4(3.3466^2)}{8}} = 2.7295$$

The test statistic is

$$T = \frac{21.8 - 10.8}{2.7295\sqrt{1/5 + 1/5}} = 6.372$$

and the p-value is 0.0002.

c. $\bar{x}_1 = 10.8, \bar{x}_2 = 21.8$ and $\bar{x} = (10.8 + 21.8)/2 = 16.3$

d. The table can be found at the end of the notes

e. The first three columns have the same sum of 163. The second three columns have the same sum of zero. Columns 7 and 8 add up to column 9 (59.6 + 302.5 = 362.1).

f. They are the same, 362.1/9 = 40.233

g. MSerror = 59.6/8 = 7.45,
   MSpaper 302.5/1 = 302.5

h. $F = 302.5/7.45 = 40.604$,
   $T^2 = 6.372^2 = 40.604$
   They are the same.

Part 3

a. $V_1 = 29.3132$, $V_2 = 22.8079$, $n = 20$, and $a = 4$

b. The test statistics are

$$MSpaper = \frac{(19)22.8079}{3} = 144.45$$

$$MSerror = \frac{(19)(29.3132 - 22.8079)}{16} = 7.725$$

$$F = \frac{144.45}{7.725} = 18.699$$

c. As the value of F is more than 18 (far more than one), we have reason to believe that these means are not equal.

d. Fcdf(18.699, ∞, 3, 16) = 0.000 02

e. We reject the null hypothesis that all means are equal, and conclude that at least one mean is different from another.

f. The sample means for 20 and 24 are close (10.8 and 13.2) so these population means could be equal. Likewise the sample means for 28 and 32 are close (20.4 and 21.8) so these population means could be equal. But all other pairings seem quite different.

Table 20.1: Sample data: Jump is measured in cm, Paper in lbs

#	Jump	Paper	#	Jump	Paper	#	Jump	Paper	#	Jump	Paper
2	10	20	18	23	32	19	22	32	15	22	28
7	16	24	17	16	32	5	14	20	9	14	24
1	10	20	8	14	24	20	24	32	13	18	28
14	18	28	16	24	32	12	19	28	4	9	20
10	13	24	3	11	20	11	25	28	6	9	24

Table 20.2: Sample calculation with sums

1	2	3	4	5	6	7	8	9
x	$\bar{x}_i$	$\bar{x}$	$x - \bar{x}_i$	$\bar{x}_i - \bar{x}$	$x - \bar{x}$	$(x - \bar{x}_i)^2$	$(\bar{x}_i - \bar{x})^2$	$(x - \bar{x})^2$
10	20	10.8	-0.8	-5.5	-6.3	0.64	30.25	39.69
10	20	10.8	-0.8	-5.5	-6.3	0.64	30.25	39.69
23	32	21.8	1.2	5.5	6.7	1.44	30.25	44.89
16	32	21.8	-5.8	5.5	-0.3	33.64	30.25	0.09
24	32	21.8	2.2	5.5	7.7	4.84	30.25	59.29
11	20	10.8	0.2	-5.5	-5.3	0.04	30.25	28.09
22	32	21.8	0.2	5.5	5.7	0.04	30.25	32.49
14	20	10.8	3.2	-5.5	-2.3	10.24	30.25	5.29
24	32	21.8	2.2	5.5	7.7	4.84	30.25	59.29
9	20	10.8	-1.8	-5.5	-7.3	3.24	30.25	53.29
163	163	163	0	0	0	59.6	302.5	362.1

Directions for folding jumping frog

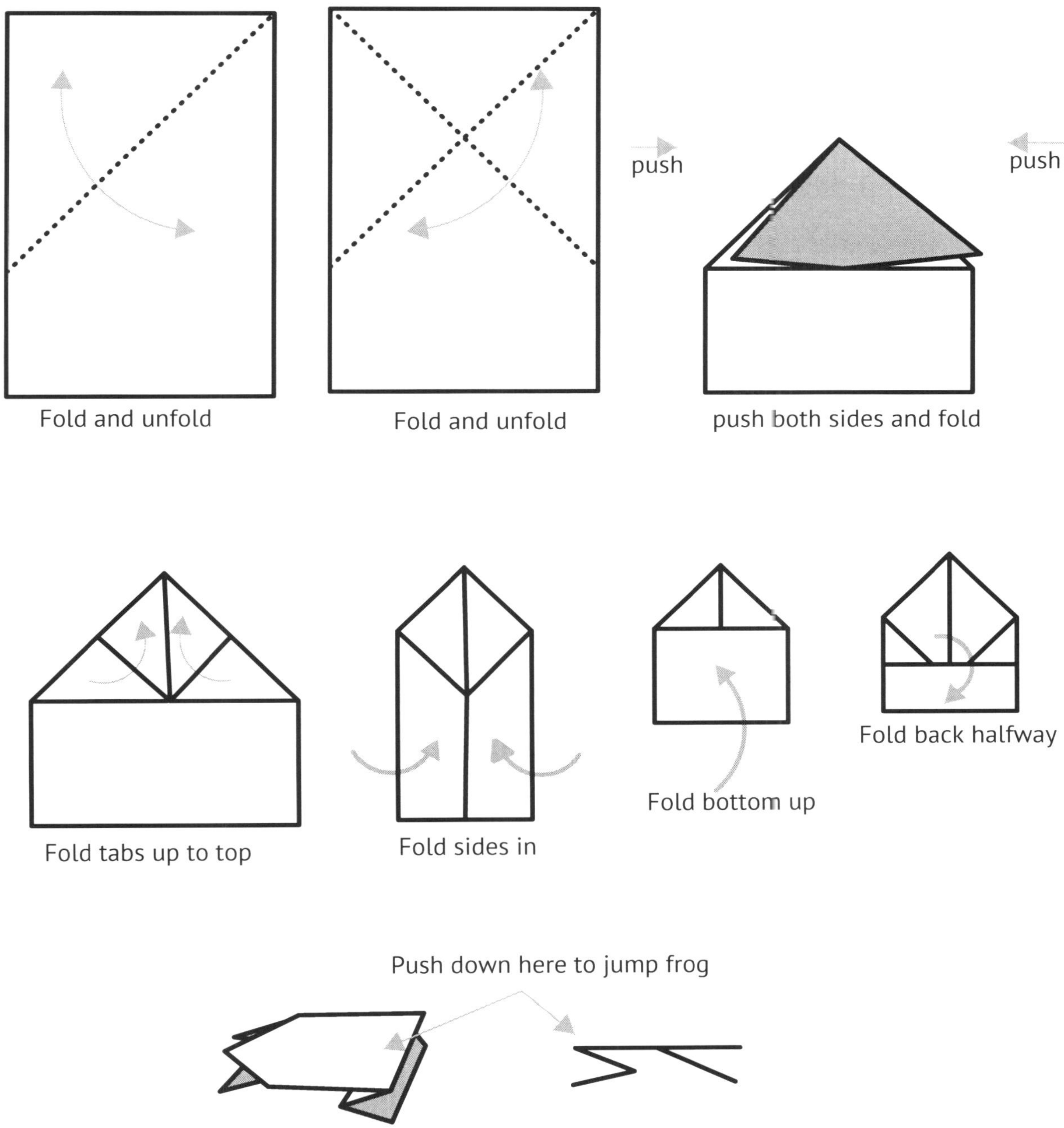

Lab 21
The Project

There are 5 deadlines in the project. Only Part 5 will be graded, but a component of that grade (10%) is the timely response to all the deadlines. As you make each submission you can expect a response back the next day. Each submission will either be approved, or returned with comments about what needs to be added or changed. Do not proceed to the next Part until you have an approval on the current Part. If a Part is not approved, you will have to submit it again with the changes recommended taken into consideration. Both the time when you first submit a Part and your final approval will be considered in your score so it is to your advantage to work early in case there are problems with your project. As a warning, most (80%) of Part 1 submissions are sent back for changes.

The deadline dates are:

0. Team Formation ______________________________
1. Project Proposal ______________________________
2. Project Design ______________________________
3. Data and Graphs ______________________________
4. Hypothesis Tests ______________________________
5. Complete Project ______________________________

Brief details of each of the Parts:

1. Project proposal will detail three things: (1) The response variables to be measured. (2) The factors to be considered. (3) Your hypotheses about the results. Note that your grade is in no way contingent on whether your hypotheses are right or wrong.

2. Project Design will detail three things: (1) The population you will sample. (2) The type of data collection, and (3) the details of the sampling. This last item should include (a) the amount of data you will collect and (b) how you will collect it including an explanation of how randomization is used in your data collection. Also (c) the units and accuracy with which you will measure your response variable should be included.

3. The Data portion will have three things: (1) The completed data set (with units of measure), listing factor levels. (2) A description of the actual data collection, including any difficulties encountered in data collection, and how the group decided to deal with the problem(s). (3) Graphs showing condition checks and what support you have for each hypothesis.

4. The statistics from your samples including sample means, standard deviations, sample sizes, least squares line, and correlation. The statistical tests (with test details) to match any and all hypotheses. Include the names of the tests, the test statistics and p-values, and the conclusion of the tests.

5. The complete project will have 5 things: The first four submissions reformatted for a report, and a conclusion summarizing your findings in relation to your hypotheses. More details and examples of this final report will be given later.

Part 1: Hypotheses

Your project will involve testing at least two hypotheses. Your team will need to define and make accurate measurements of some quantitative response variable. For this Part you need only worry about creating the hypotheses you will be testing. This Part will be approved providing the project is neither too hard nor too easy for you to get a good grade. Many proposals will be rejected (or sent back for improvements) because of things you don't yet know or understand so don't feel bad if you are rejected, just plan on it.
You must decide on your quantitative response variable, and both a quantitative and qualitative factor that you believe has some influence on your response. You must then propose two hypotheses about this influence. You need to use the words, response and factor, properly in your proposal. The values of the factors may be set or measured. Here are some examples to explain this:

a. The response measured in this project is the final score earned playing a particular computer game. The qualitative factor is whether or not the player has been taught a special trick to playing the game. The quantitative factor is the score on the game before teaching the trick. The first hypothesis is that those with higher scores the first time they play will also be the ones that have the higher scores the second time. (There is positive association between first score and second score.) The second hypothesis is that the mean improved score for those who are taught the trick is higher that the mean improved score for those that are not taught the trick.
In this example the numeric factor (score first time) is measured and the categoric factor (teach or don't teach) is set or assigned.

b. The response measured in this project is the voltage of batteries after 8 hours of continuous use in a flashlight. The qualitative factor is the brand of battery. The levels on this first factor will be Duracell and Energizer. The quantitative factor is the temperature that the flashlight is kept at during the 8 hours. The first hypothesis is that after 8 hours Duracell batteries have a lower mean voltage than Energizer. The second hypothesis is that flashlights that are kept colder have higher voltages after 8 hours. (There is negative association between temperature and final voltage.)
In this example the numeric factor (temperature) is set or assigned, and the categoric factor (battery brand) is measured.

Part 2: Design

Deadline: ____________________

You must now plan the collection of data. Once this is approved you can begin data collection. Don't start getting data until your design is approved.

Population

In this Part of the project you need to first identify the population(s). Some of the projects involve human subjects. The bigger your population, the better your conclusion can be, but it makes it harder to collect the data. If your population were all students in the world then you would have to find a way to get volunteers from that entire population. If your population is only students at this school, this will make it easier to gather a sample, but your conclusion can only be about students here, not all students. If your subjects are only friends or only the members of the project team, then your conclusions ...

If the project involves items purchased, then the conclusion must still reflect your population. "Items purchased in year 20XX in the city of XX available at the XX store." For a more general conclusion consider expanding your population as widely as you reasonably can.

Randomization

The key here is that randomization is critical to your design. No design will be accepted if it does not explain how the randomization is implemented. Your task is to explain how you will collect your sample and how much data you plan to collect. Give the units of measure and the accuracy you will use.

Examples

Here are two examples of what this submission should look like.

a. The hypotheses are that those with higher scores in the first round of playing the video game will also be the ones that have the higher scores in the second round, and that the mean improved score for those that are taught the trick to the game is higher that the mean improved score for those that are not taught the trick.

   The population will include volunteer students from this school with a full range of gaming experience. We will recruit about 40 people making sure that we have some with little or no experience at gaming as well as some who play such games daily.

   A round consists of two plays at the game with the higher of the two scores recorded. We will then rank the scores from low to high. Starting with players number 1 and 2 in the ranking we will flip a coin. If it lands heads we explain our strategy of game to the odd number (#1) and give some meaningless tip to the even number (#2). If the coin lands tails these are reversed. These two players then play another round of the game and again we record the higher score. This is repeated for players ranking 3 and 4 and so on to the end of the list. At the end we will have 20 who have learned the trick and 20 who did not.

b. The hypotheses are that Duracell batteries will have a lower mean voltage than Energizer after 8 hours of use, and the lower the temperature of use the higher the voltage after 8 hours.

   The population will include 5 four packs of Duracell and Energizer batteries purchased from 5 different locations in town. They will be randomly selected from what is available at the store. We will also purchase two identical flashlights and label these A and B.

   We have for our use one small refrigerator and one small electric oven. Using a random number generator we will pick a random value between 0 and 200 and adjust either the refrigerator, the freezer, or the oven to that temperature. Then taking two Duracell batteries we will flip a coin. On heads we will load these into flashlight A and tails into flashlight B. We will then load two Energizer batteries into the remaining flashlight. Next we will turn both lights on and place them in the temperature controlled environment. After 8 hours we will remove the flashlights and then test each battery with a digital voltmeter recording the voltage to 0.001 of a volt for all four batteries. Then we will start the whole process again until we have 20 readings for each type of battery at 10 different temperatures.

As before, if your submission lacks sufficient detail or is incorrect, you will have to make corrections and submit again. This could take two or three attempts so make your first submission in advance of the deadline to ensure that you have time to fix it correctly.

Submission should be in this format:

- The two hypotheses
- The population
- The design (including randomization)

Part 3: Data

Deadline: ______________________

The standard format for data is to have a single table for all of the data. These values should appear in the order they were collected. The first column contains values of the response variable. Additional columns are used to record the data associated with this measure. You will need a column for each of your factors, and perhaps additional columns for when, where, or who collected this piece of information.

Voltage	Brand	Flashlight	Temperature	Date / time	Measured by
1.108 V	Energizer	A	47 °F	11/5 5:10 PM	Sue
1.021 V	Energizer	A	47 °F	11/5 5:10 PM	Sue
0.754 V	Duracell	B	47 °F	11/5 5:10 PM	Sue
0.831 V	Duracell	B	47 °F	11/5 5:10 PM	Sue
0.662 V	Duracell	A	103 °F	11/6 4:58 PM	Rick

In addition to the data table you need to produce graphs that show support (or no support) for your hypotheses as well as those which check conditions for accurate testing. Each graph needs to be well labeled and carefully made. You are not required to use a computer to make these plots but it is strongly recommended.

Categoric Factor

For each value of the categorical variable, make a separate list of the values of the response variable. For each of these lists you must make a histogram and a time series plot. You must also make a parallel box plot with all lists on the same plot. If you have made your categorical variable dichotomous, (binomial) then you will have 5 plots (2 histograms, 2 time series, and 1 parallel boxplot).

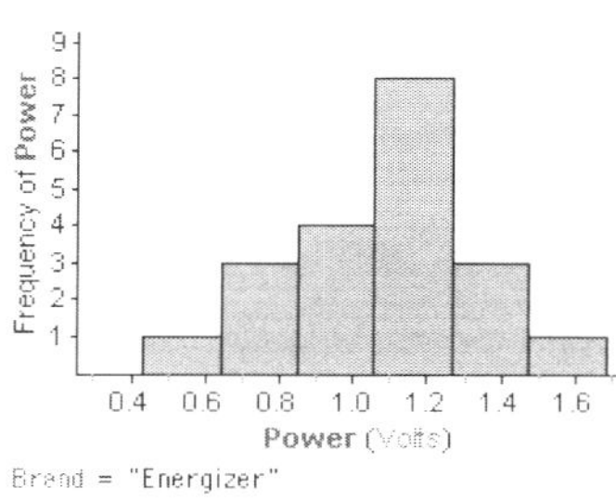

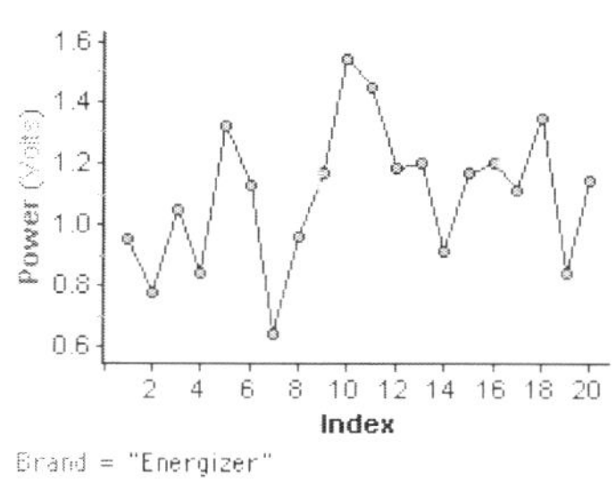

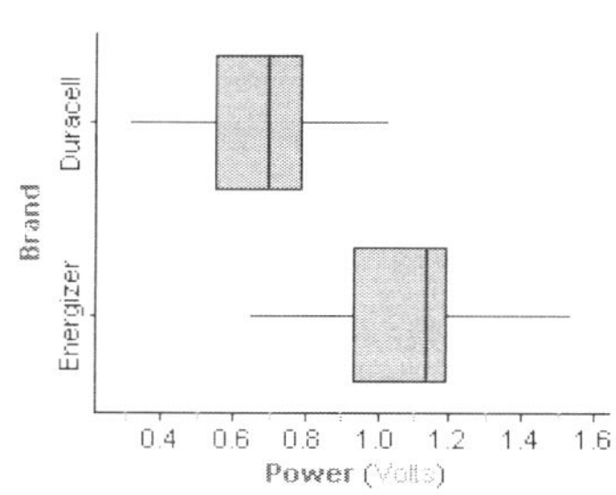

Numeric Factor

For the numeric variable you need a scatter plot of the ordered pairs (factor, response). If you see a need, you may want to do separate scatterplots for the different values of the categoric variable. For example, if the voltages for the two brands of battery were very similar, then one plot would be right, but if they were quite different then you would make a separate scatterplot for each brand of battery. Add the line of best fit to the plot(s). Then create a residual plot of the errors in that model. Finally create a time series plot of the residuals.

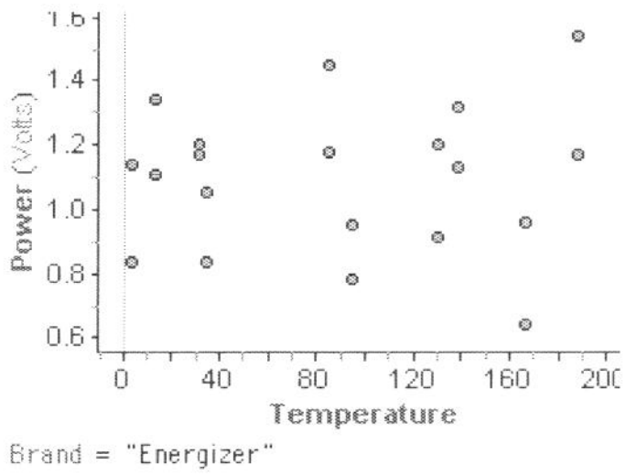

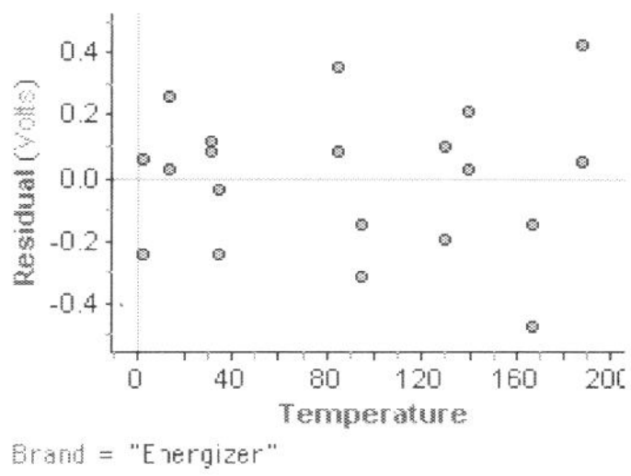

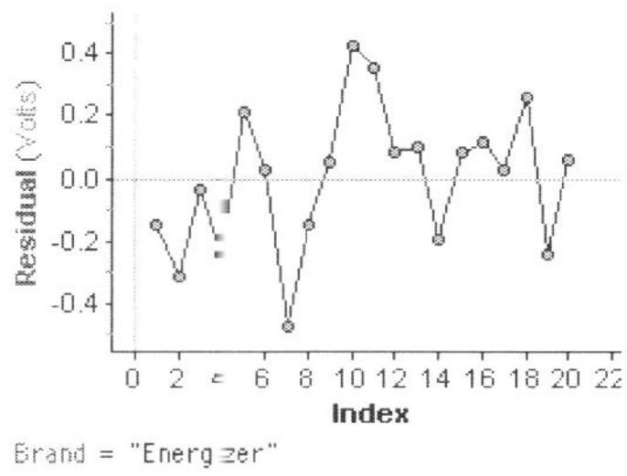

Not due at this deadline but needed for your final report: Each graph needs a short but complete comment. The comment about each histogram should focus on the shape and outliers of the distribution of that sample, and if conditions for the test are properly met. The time series plots should have comments about patterns and independence. The comment about the box plots should say if there is graphical support for the hypothesis or not. The comment about the scatter plot should indicate if there is a "trend" and how strongly the graph shows that trend.

Part 4: Testing

Deadline: ______________________

This is the testing of each hypothesis. Depending on how many values you have for your categoric variable, and if you decided to split the data for the least squares, you MAY now have more than two hypotheses to test. You should record these things:

a. Summary statistics: This will depend on the hypothesis. It may include the means, the standard deviations, and the sample sizes. Or it may include the least squares line, the correlation, and the root mean square error.

b. Identify which test is appropriate for your data.

c. Give the symbolic statement of your null and alternate hypotheses.

d. Identify the significance level of your test.

e. Give the value of the test statistic.

f. State the degrees of freedom.

g. State the p-value.

h. State the symbolic conclusion and justification.

You should already have created the graphs needed for any condition checks. If not, you will need to add these to the list. Now do it all over again for the other hypothesis.

Example

Statistic	Energizer	Duracell
n	20	20
$\bar{x}$	1.096 V	0.6810 V
s	0.2283 V	0.1636 V

Statistic	Value
LSL	$V = 1.079 + 0.0001894T$
Corr	$r = 0.0536$
RMSE	$s_e = 0.2342V$

Two Sample Pooled T-Test

H_0	$\mu_E = \mu_D$
H_A	$\mu_E > \mu_D$
α	0.05
T	6.167
df	38
p-value	< 0.000

Reject H_0

$0 < 0.05$

Linear Regression T-test

H_0	$\beta = 0$
H_A	$\beta < 0$
α	0.05
T	0.08995
df	18
p-value	0.54

Fail to reject H_0

$0.54 > 0.05$

Part 5: Final

Deadline: ______________________

Create a title page giving the critical information about your project. Next assemble the first 4 Parts of the project into a single document. Clean up the tenses because they are all in the past now. Keep the order of the Parts, but feel free to add transition sentences to make it read better. You need to include the complete data set in the order it was collected. This can be part of the report, or you can place it in an appendix. You also must submit the data alone (in the proper format) electronically either through e-mail or on a flash drive. The format should be .csv, .txt, .xls, or .doc

Graph comments

If you have not yet written a comment of for EACH graph then you need to do that now. The comment about each histogram should focus on the shape and outliers of the distribution of that sample, and if conditions for the test are properly met. The time series plots and residual plot should have comments about patterns and independence. The comment about the box plots should say if there is graphical support for the hypothesis or not. The comment about the scatter plot should indicate if there is a "trend" and how strongly the graph shows that trend.

Conclusion

After all the Parts are in, then you need a concluding paragraph. This will provide the conclusions to your hypotheses in context, any concerns you have about the accuracy of the tests, and any warnings you have about the limitations of your results (especially pertaining to the population of study).

Grading

The next page is the rubric used when grading the project. Please use it as a checklist to make sure that you have included all of the needed parts. There is also a sample report for you to read and model.

Every member of the team should read through the entire report, not just the parts they wrote. Note that if your name is on the report then you are responsible for all errors in it. If some member of the team is unwilling to contribute to the report and will not help in proof-reading the document, then you should not put their name on the report. As in real life, you get credit for what you do and blame for what others do in your name.

Statistics Project Grading Rubric

1. Did the title page have a title for the project,
   the course name, the date,
   the team name, and the names of the members?

2. Did the opening identify the response variable,
   all factors and the levels of the factors to be considered?
   Were the terms response, factor, and level used correctly?

3. Did the opening clearly state the two hypotheses in words?
   Did it include the idea of population means?
   Was the significance level for the test stated?

4. Was the design for data collection clearly stated?
   Was the population defined?
   Were there details of how the data collection went,
   what problems were encountered,
   and how they were resolved?

5. Was a complete list of all the data, with units, included
   in the proper format?
   Was it in the order the data was collected?
   Was an electronic file submitted?

6. Were all graphs included?
   Were the graphs properly titled and labeled, including units?
   Was there a comment about each graph?
   Was the comment related to either the test accuracy or the hypothesis?

7. Were there summary statistics for the data?
   Were the values correct?
   Were the units correct?
   Was there a comment connecting these to the hypothesis?

8. Were the correct types of tests identified for each hypothesis?
   Was there an explanation of why that test was chosen?
   Did the test done match the named test?

9. For each test: Was the test statistic given?
   Were the degrees of freedom given?
   Was the p-value given?
   Were the values all correct?

10. Was the conclusion given symbolically
    and in context?
    Was justification for the conclusion stated?
    Was it related back to the population being tested?
    Were concerns with the accuracy of the test brought up?

Statistics Project

A Study of Batteries

May 2016

Team Omicron Members
Rick Brandon
Clarence Braun
Sue Beauregard
Steve Flores

A Statistical Study of Batteries

In this study we look at the voltage of batteries. The batteries were 1.5 volt D-cell purchased new and placed into a flashlight which was then left on for 8 hours. Our response variable was the voltage of the battery after eight hours. We tested two hypotheses in this study. The first factor was the brand of battery. The team felt that Bunny Batteries brand was better than Copper Top brand so we created the hypothesis that the mean power after 8 hours would be higher for the Bunny batteries than the mean power for Copper batteries. The second factor was the operational temperature of the battery during use. The team believed that when cold, the batteries would keep more of the charge than when hot. We created the hypothesis that there is negative association between temperature and the charge remaining after 8 hours.

Our population included any batteries that could be purchased in Traverse City this year. We randomly selected 5 stores in the area that we knew sold batteries. From each store we randomly selected one 4 pack of Bunny and one 4 pack of Copper batteries. This gave us 40 batteries to use for our study. Our design also included the use of two new identical metal flashlights, labeled A and B, one mini-refrigerator, one small electric oven, and one digital voltmeter. Each day we used random numbers to choose two of the remaining Bunny batteries and two Copper batteries. Using a coin flip we loaded the two Bunny batteries into flashlight A or B and the two Copper batteries into the other flashlight. We then used a random number generator to pick a temperature between 0 and 200 °F. When the temperature was above room temperature we used the oven that was set to that temperature. When the value was below room temperature we had to guess the coldness setting that would give us about the right temperature in the refrigerator. We placed a thermometer in the refrigerator to tell us what the true temperature was. Both flashlights were turned on at the same time, and the time of day was carefully recorded. Then the flashlights were placed in the oven or refrigerator. After 8 hours the flashlights were removed and turned off. The batteries were then removed from the flashlights and measured using the voltmeter. These values were then recorded into the data table. (See table 1). This process was repeated for 10 days over three weeks until we had used all 40 batteries.

Data collection went mostly as planned. The clear plastic lens on the flashlight did some melting at the higher oven temperatures but this did not seem to affect the working of the flashlight so we ignored it. The temperatures in the refrigerator were not always very close to the number that was given by the random number generator as we were guessing on where to set the dial, but we still counted it as good data and used the temperature from the thermometer.

Our first hypothesis was that the mean power after 8 hours would be higher for the Bunny batteries than the mean power for Copper batteries. Now if temperature was a real factor in the power then we would treat this as a matched pair test to remove the influence of the temperature on the power. But in this case we found that this is not true (see later in the report), so we treated this data as two independent samples of size 20.

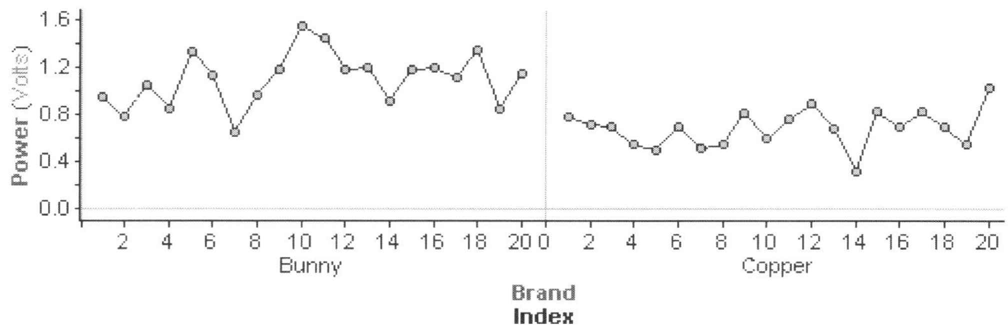

The time series plots show no real pattern so we conclude that the power measures were independent of time. Because we used randomization in our selection of batteries we feel that we have met the independence condition for an accurate T-test.

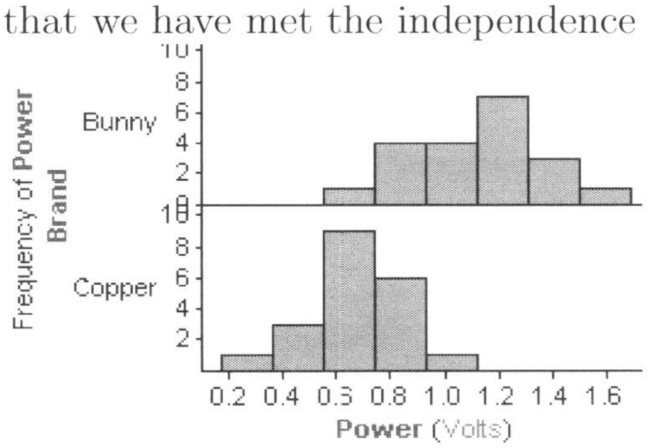

The histograms show mounded and mostly symmetric data for our samples. Together with a sample size of 20 we feel that this gives us reason to feel that our sample means will be from approximately normal distributions. So the p-value from our test should be accurate.

The parallel box plots show real support that the center for Bunny is higher than the center for Copper.

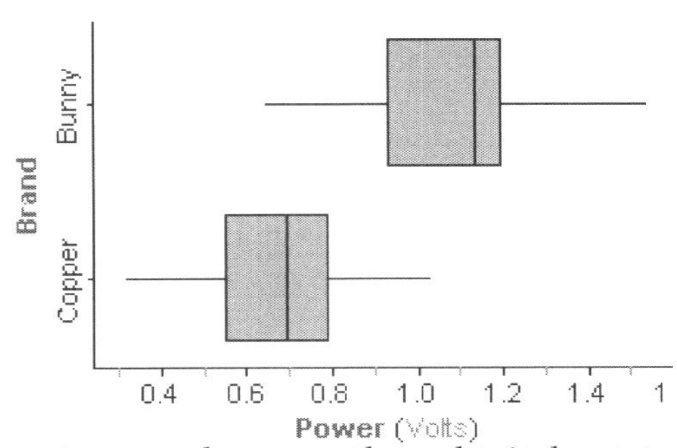

Statistic	Bunny	Copper
n	20	20
$\bar{x}$	1.096 V	0.6810 V
s	0.2283 V	0.1636 V

Because our two sample standard deviations were not very close we thought it best to be conservative and do a two sample T-test with un-pooled variances.

$$H_0 : \mu_B = \mu_C$$
$$H_A : \mu_B > \mu_C$$
$$\alpha = 0.05$$
$$T = 6.607$$
$$df = 34.44$$
$$p\text{-value} < 0.0001$$

Because our p-value is much less then our significance level we reject the null hypothesis and conclude that the mean of Bunny Batteries is greater than that of Copper Top after 8 hours of continuous use.

Our second hypothesis is that the power is negatively associated with the temperature of the battery during 8 hours of use. Now if Brand were not a factor in the power, then we planned to treat this as one sample of 40 batteries. But in this case we found that this is not true (see earlier in the report) so we separated the data into two independent samples of size 20 to remove the influence of the brand on the power.

Bunny Batteries

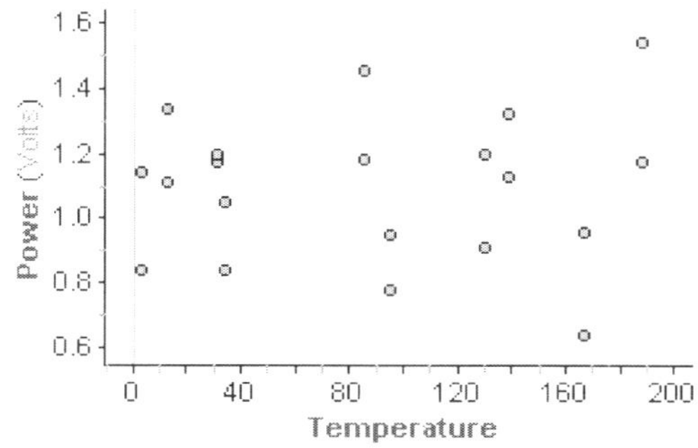

There is almost no pattern or trend to the scatter plot of temperature vs. power. So there is not much support for the stated hypothesis.

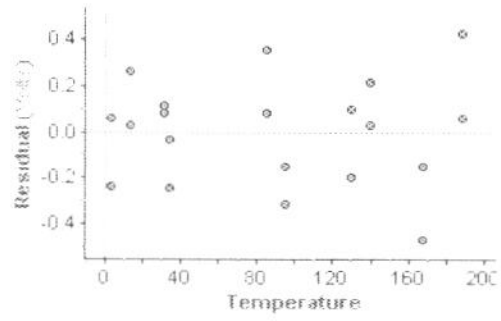

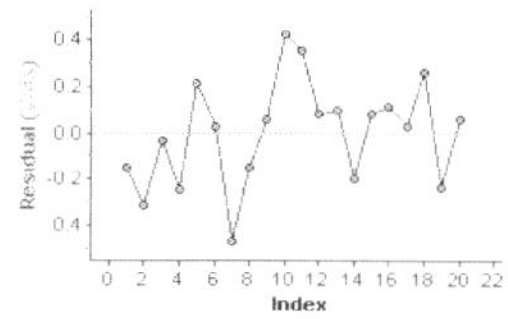

The residual plot and time series plots show no real pattern so we conclude that the power measures were independent of time. Because we used randomization in our selection of temperature and batteries we feel that we have met the independence condition for an accurate T-test.

$$H_0 : \beta_1 = 0$$
$$H_A : \beta_1 < 0$$
$$\alpha = 0.05$$
$$T = 0.2278$$
$$df = 18$$
$$p\text{-value} = 0.59$$

Because our p-value is much greater then our significance level we fail reject the null hypothesis and cannot conclude that the running temperature is associated with the voltage of Bunny Batteries.

Copper Tops

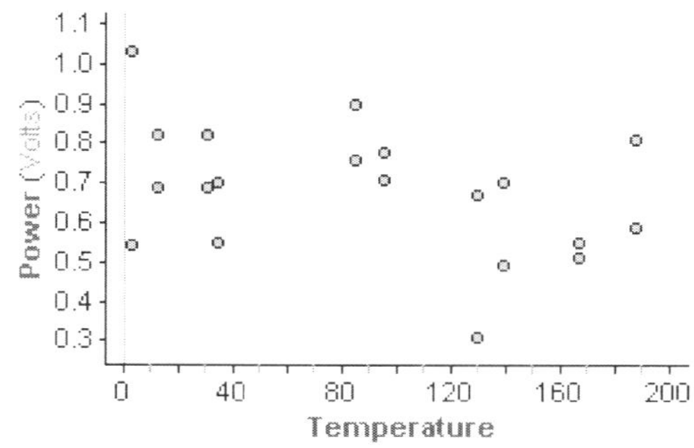

There is slight downward trend to the scatter plot of temperature vs. power. So there is some support for the stated hypothesis.

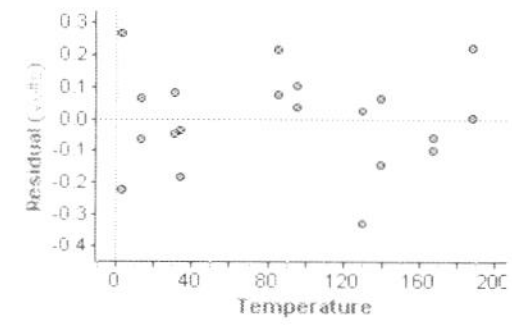

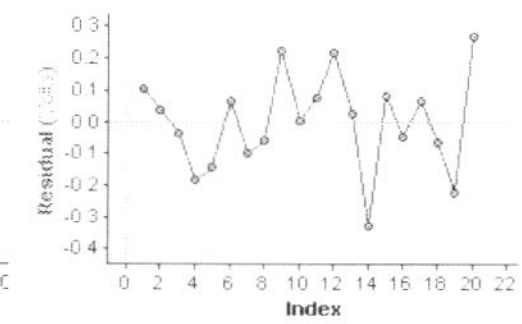

The residual plot and time series plots show no real pattern so we conclude that the power measures were independent of time. Because we used randomization in our selection of temperature and batteries we feel that we have met the independence condition for an accurate T-test.

$$H_0 : \beta_1 = 0, H_A : \beta_1 < 0$$
$$\alpha = 0.05$$
$$T = -1.735$$
$$df = 18$$
$$p\text{-value} = 0.05$$

Because our p-value is slightly less than our significance level, we reject the null hypothesis and can conclude that the running temperature is negatively associated with the voltage of Copper Tops.

In conclusion we have found that Bunny batteries bought in Michigan in 2015 have a higher mean voltage after 8 hours of continuous use than the mean voltage of Copper Tops. This may or may not apply to other locations, other years, or other load durations. The p-value showed a highly significant value and the conditions were met for an accurate test. We also found that Bunny Batteries energy use was not associated (or dependent on) the temperature to any significant level. On the other hand we found that energy use was higher at higher temperatures for Copper Tops. This association was weak but did show that 14% of variation in energy was do to changing operational temperature. Again this was our finding for Traverse City purchased batteries in 2015, and might not apply to other locations or times.

Table 21.1: Data in the order it was measured

Power (V)	Battery Brand	Temperature (°F)	Flashlight	Power (V)	Brand	Temperature (°F)	Flashlight
0.95	Bunny	95	A	1.45	Bunny	85	A
0.78	Bunny	95	A	1.18	Bunny	85	A
0.78	Copper	95	B	0.76	Copper	85	B
0.71	Copper	95	B	0.9	Copper	85	B
0.7	Copper	34	A	0.67	Copper	130	A
0.55	Copper	34	A	0.31	Copper	130	A
1.05	Bunny	34	B	1.2	Bunny	130	B
0.84	Bunny	34	B	0.91	Bunny	130	B
0.49	Copper	139	A	1.17	Bunny	31	A
0.7	Copper	139	A	1.2	Bunny	31	A
1.32	Bunny	139	B	0.82	Copper	31	B
1.13	Bunny	139	B	0.69	Copper	31	B
0.64	Bunny	167	A	1.11	Bunny	13	A
0.96	Bunny	167	A	1.34	Bunny	13	A
0.51	Copper	167	B	0.82	Copper	13	B
0.55	Copper	167	B	0.69	Copper	13	B
1.17	Bunny	188	A	0.54	Copper	3	A
1.54	Bunny	188	A	1.03	Copper	3	A
0.81	Copper	188	B	0.84	Bunny	3	B
0.59	Copper	188	B	1.14	Bunny	3	B

Teacher Notes for The Project

Prerequisite Knowledge

- Design (by Part 2)
- Graphs and Stats (by Part 3)
- Two sample means test (by Part 4)
- Linear regression test (by Part 4)

Objectives

The objective of this lab is to bring the separate parts of sampling and analysis together in a single project. While it is possible to squeeze this activity into seven weeks, We prefer to assign the project in the first week of class and complete it in the last week of class. We only hand out the activity one Part at a time. Thus, students don't get Part 2 until they have completed Part 1. As teams need to be formed in the first week before we know the students, we usually have them help by asking them to sit with people they like to work with. Then we work our way around the room, making groups of 4 or 5 students. The groups then choose a group name. Usually we provide context. Some years it's pick a greek letter, others its pick a color, or pick a vegetable. This allows us to refer to a group easily later on.
As this is a long term project you need to give groups structure and assistance. In our classes this project grade is equal to any test grade. The groups might elect a leader to take responsibility for knowing who is doing what and keeping track of all the timelines. Assure the groups that they are always welcome to talk to you about the project either as a group or individually. Have them tell you if things are not going smoothly. We don't like to change the groups once the project has started, but have done so to help keep things moving. Keeping a chart of the date each group has gotten approval for each Part and periodically showing this chart to the class to remind them of the current deadline can be helpful.

Comments and Answers

Part 1

The groups always seem to need a lot of help here. We say "pick something you measure, not something you count" and then they typically say "The number of people ... " After a couple of go arounds they get it figured out. We always make them get this Step approved. Some groups will take on a project way too big or too hard and we need to bring it down. Others will do something too simple "Guys are taller than girls", to which we reply "You want to spend this whole course figuring out something you already know?" Tell students not to worry too much about how they will collect the data at this point, that will come later, all they need is the hypotheses. So it is up to you to be thinking ahead about how they will do it and if they can do it.

Part 2

The goal is to be very thorough in this design. Don't accept "randomly choose". You want to know if this will involve a coin, a die, a blindfold, a calculator, or what. In general, what is the means of randomization? If there are survey questions then you will want to read them. If possible, ask them to try their design at least once to find how well it works. If using a tool to measure, think about what kind of accuracy this tool provides. In the past we have had groups measure to the nearest inch only to find that all their measures were one of three values. They may not yet have discussed in class what is an appropriate sample size so you may need to help them determine how much data they should be going for.

Part 3

To keep things moving don't make them redo this part if there are small errors. Just list all the errors such as missing labels or units on graphs, and tell them that these need to be fixed before the final report. Do not check to see if the graphs are correct at this time. Students should be warned that approval does not mean that everything is correct, only that they have done all the things requested.

Occasionally a group will not have completed all of data collection by this deadline because it is taking longer than they planned. In this case they should make the graphs with the data they have in order to meet the deadline. At this point we have, on rare occasions, found that a group differed from what we thought was the design and have collected data that can't be used for the tests we wanted them to do. In some cases we have them recollect the data (if can be done in a short time) or find some other way to use the data. These groups can end up with proportion tests, chi squared tests, or even ANOVA tests depending on the data. Hopefully this won't happen to your groups but we wanted you to be prepared.

Part 4

You do not need to check the calculations on the tests at this point, just make sure they are listing the name of the test, the test statistic, the degrees of freedom, and the p-value. Remind them that they are responsible for checking their own work and approval on this Part does not mean that the numbers are right, just that they have done the right tests.

Part 5

In assigning this Part we tell them to use the grading rubric as a check-list for their final report. We often have some old reports, with the cover page and names removed, and a graded rubric for them to read if they wish. We find that most groups do not take the time to look at old reports but they do look at the sample of a "perfect" report that we provide. We have always allowed groups to submit all or partial reports early for us to comment and correct. However, this option ceases in the 24 hours before the report is due. We always ask groups "Who did the calculations?". We then ask "Who checked their work?". The same with the graphs and the writing of the report. You will have to decide how to give points on the rubric. What we have used is ...

- 10% Deadlines
- 5% Title page
- 7% Description
- 8% Hypotheses
- 10% Design
- 7% Data
- 18% Graphs
- 10% Stats
- 7% Test
- 8% Test values
- 10% Conclusion